CLASSIC GEOLOGY IN EUROPE 2

Auvergne

CLASSIC GEOLOGY IN EUROPE 2

Auvergne

Peter J. Cattermole

TERRA

First published in 2001 by Terra Publishing

Terra Publishing
PO Box 315, Harpenden, Hertfordshire AL5 2ZD, England
Telephone: +44 (0)1582 762413
Fax: +44 (0)870 055 8105
Website: http://www.terrapublishing.net
E-mail: publishing@rjpc.demon.co.uk

ISBN: 1-903544-05-X paperback

10 09 08 07 06 05 04 03 02 01 00
10 9 8 7 6 5 4 3 2 1

British Library Cataloguing-in-Publication Data
A CIP record for this book is available from the British Library

Library of Congress Cataloging-in-Publication Data are available

Typeset in Palatino and Helvetica
Printed and bound by Biddles Ltd, Guildford and King's Lynn, England

iv

Contents

Preface

Since first visiting Auvergne with my family over 20 years ago, I have been regularly drawn back. It is an area that immediately captivated me on account of its superb lush scenery and the striking way in which its volcanic landforms presented themselves. Over the ensuing years I have taken many students to these parts as well as leading commercial tours for people with but a general interest in volcanoes and scenery. Few appear to have been disappointed, indeed, most have been not only impressed but also excited by what they have seen.

As a volcanologist I am naturally attracted to areas where volcanoes are currently active. None of Auvergne's volcanoes are likely to erupt again, but the famous puys and their associated deposits are so well preserved, so accessible, and so concentrated in a relatively small area, that one can quite vividly imagine how it was, just a few thousand years ago, when this part of France was in the grip of a violent and unforgiving Earth.

Today Auvergne is tranquil and easier to explore than it once was. New or improved roads and marked footpaths make it readily accessible to anyone interested in geology, or to those who simply enjoy walking amid stunning scenery. This guide is offered to all those who seek to understand something of the geology behind this unique landscape.

Although this guide is based on my own explorations of the area, I have drawn on the work of several French geologists for some of the illustrations and also have used information displayed within the Parc des Volcans; these sources are acknowledged within the text. I would also like to thank all those who have offered their support during the preparation of this book, in particular my publisher, Roger Jones, with whom I have explored in words not only the geology of Auvergne but also the intricacies of the French language. With regard to the latter, I should mention that, during a recent visit to the region, three different maps and three different road signs gave five different spellings of the name of one village! I trust that the reader will therefore understand if such variance is found within the book.

Finally, I dedicate this book to Jack, navigatrice extraordinaire.

Peter J. Cattermole
Sheffield
August 2001

Frontispiece A postcard scene from 1907: travellers arriving by tram at la Baraque, with Puy de Dôme in the background.

504 *Arrivée du Tramway à la Baraque et le Puy-de-Dôme*

Introduction

Fifty years ago, before the motor car had become a universal people mover, many outsiders would have had extreme difficulty in pinpointing the location of Auvergne. This would have applied as much to the French as to the British and other foreigners. Indeed, until quite recently French government considered Auvergne to be a difficult and backward region and, in consequence, it traditionally enjoyed relatively little political and fiscal support. Fortunately, things have now changed. Auvergne has been opened up by the building of new roads and upgrading of old ones, by inventive marketing of tourism and via the allocation of resources that have provided facilities previously unheard of, including visitor centres, excellent information boards and well equipped leisure areas – tourists have now discovered this hitherto unfashionable region. The changes wrought have been to the advantage of visitor and Auvergnat (a native of Auvergne) alike: the former is now able to explore a most marvellous region and the latter now has a source of income in addition to agriculture, the traditional means of living, which has always been arduous in difficult terrain and often very testing weather.

The geographical centre of France is on one of the routes entering our region from the north. Here, between Bourges and Montluçon, lies Bruère-Allichamps, where a Roman milestone is supposed to mark the spot. Modern surveying suggests that the modern "centre" is at Saulzais-le-Potier, some 25 km farther south. Either way, Auvergne is a part of the Massif Central, a huge upland region that is geologically ancient and which provides a physiographical barrier between the lowlands of northern France and the Mediterranean coast. It consists of four départements: Allier, Puy-de-Dôme, Cantal and Haute-Loire, which together cover an area of 55 000 km². Much tourist information focuses on the marvellous Parc des Volcans, the region characterized by France's largest concentration of recently active volcanoes. This is hardly surprising, since its scenery and geology are quite mesmeric and, indeed, they provide the main subject matter for this guide. However, there is much more to the region than this: there are also wide and fertile plains, extensive forests,

1

Figure I.1 The village of Roussac in the Cantal. Village life goes on here in a landscape of high granite cliffs capped with basalt.

deep ravines, U-shape glaciated valleys, sharp arêtes and rolling hillsides moulded by ice sheets. When you superimpose on this the amazing variety of buildings, farmsteads and the diverse patterns of agriculture that this varied terrain has demanded of humans trying to farm it, you have a region characterized by immense diversity rather than regional unity.

Auvergne is a paradise for the geologist, naturalist and walker. Although it is a large region and there are many tracts far from roads, you do not have to walk miles to study the rocks. In general, the geology is readily accessible, there being many quarries and roadside exposures, and many of the loftiest spots can be reached by road or cable car. However, those willing to tread the more remote byways – the region is magnificently threaded with footpaths – can discover localities where they will seldom be troubled by fellow travellers (Fig. I.1).

The many towns and villages of the region are charming and the spring flowers wonderful. Indeed, flowers are to be found throughout the summer months too. Soaring birds of prey are almost constant companions to the traveller in Auvergne, as are the handsome cattle, particularly the Charollais. The pattern of transhumance that has been practised for generations has left isolated burons (a distinctive kind of stone barn) scattered over the summer pastures, where cheeses were made on site. Auvergne really is a magnificent part of the world in which to study geology, and I hope that this guide will encourage and enable my readers to enjoy its many facets to the full.

Chapter 1

Auvergne – an island surrounded by land

Auvergne has been described as an island surrounded by land, an apt description, for the terrain is different, the people are different, its history is different, and the weather is different from the surrounding regions. It is a wild part of France, a gnarled but beautiful upland region moulded by volcanism, wind, water and ice. On the geological timescale, the volcanic upheavals that threw up the famous puys and erupted rivers of molten lava over the granite basement occurred only yesterday. Indeed many of Auvergne's volcanoes were still erupting when the glaciers of the most recent ice age were still moving remorselessly towards the surrounding lowlands. At this time, central France was a chain of fire, with dozens of volcanoes spewing out ash and dust, spawning molten lava floods and, more dangerously, erupting glowing avalanches, or **pyroclastic flows**.

The landscape has a majestic serenity. The higher volcanic centres have been much modified by glaciation, incised by countless streams and adorned with sparkling waterfalls. The less lofty volcanic cones have been colonized by trees and shrubs, or planted with pine and spruce, and many craters have been filled with lakes. Several areas, once barren rocky wastes, have become lush green valleys, but there are still barren crags, ravines and underground lava tunnels. Furthermore, human beings have modified the landscape, building barrages that have dammed several of the principal rivers, creating large artificial lakes, often of startling beauty.

The island identity of Auvergne is further reinforced by the fact that the old granitic basement underlying the region is bounded by several major faults that trend either northeast to southwest, or slightly east of a north–south line. The one bounding the plateau to the east is particularly obvious, since it forms the western wall of the wide Allier Valley, in which the important city of Clermont-Ferrand now sits. On the opposite side of this valley, another fracture defines the western edge of the Monts du Forez. Although much of the observable fracturing may be relatively recent in geological terms, it follows a much older zone of weakness that continues

3

northwards and is allied to an **en echelon** zone of fracturing that passes through the region of Bresse (north of Lyon) and eventually into the Rhine Rift Valley. The visitor to the northern part of the region – Puy-de-Dôme – will have little difficulty in imagining what Auvergne was like at the time the volcanoes were active. A great line of cones runs north–south as a kind of spine, and flat-topped floods and blocky fingers of lava can be seen fanning out towards the lower ground on either side (Fig. 1.1). On a clear day, the highest of these volcanoes, Puy de Dôme, the only single mountain to give its name to a département, can be seen like a massive beacon from almost anywhere that is raised above the plateau. The huge mast that now crowns its summit adds a bizarre touch of human identity, quite different from the modest Roman temple of Mercury to be found there.

Travelling farther south, volcanoes are not so conspicuous, although some still stand proud, and the geological history has to be pieced together from structures and deposits associated with the much more deeply eroded remnants of two huge **stratovolcanoes**, those of Monts-Dore and

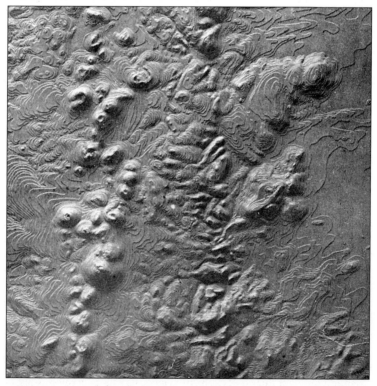

Figure 1.1   The puys and lava flows of Puy-de-Dôme as seen from the air. Taken from an old relief plan attributed to Captain Tallon and published around 1900.

Figure 1.2 The glaciated peaks of the Cantal, viewed from the summit of Puy Mary.

the Cantal. Both are older than the volcanoes of the north and, because they were much loftier, have been extensively glaciated; their greater degree of dissection calls for ingenuity to discern what went on there (Fig. 1.2). Fortunately, a series of beautiful radiating valleys provide sections through these old volcanoes, and lead to their summits, where naked peaks stand proud of grassy surroundings. The evolution of these two great volcanic centres can be traced by following itineraries that elucidate their stories. Because of their altitude, the slopes of Sancy/Monts-Dore and Cantal provide one of the finest cross-country skiing areas in France. Tows and lifts are abundant, some operating throughout the year, providing the summer visitor with easy access to some of the higher summits. It is strange to think that the smooth snowy paths now followed by recreational skiers were once channels for incandescent flows of ash, inimical to snow and ice. Heed one warning, however: even in July and August the weather can change dramatically. Snow can fall in September, and heatwaves may occur as early as April. Be prepared for almost anything.

Other popular recreational facilities are provided in the region. Puy de Dôme itself has become the headquarters for European hang gliding; then there is canoeing and rafting, rock climbing and even some pot holing and caving. The fishing is superb and, for the more adventurous motorist and cyclist, there are narrow winding backroads that lead to countless attractive hamlets hidden away in the hills, and remote village inns and excellent restaurants. Paradoxically, Auvergne is also a haven for those who are unable to be so active. There are many spas and thermal springs – known since Roman times – where those with less than perfect health can come and bathe their troubled bodies or drink the mineralized waters. Vichy, le Mont-Dore, la Bourboule and St-Nectaire are but four of these centres. For those who enjoy history or architecture, some of Europe's finest Romanesque churches are to be found here, and there are many fine old buildings

5

Figure 1.3 The hill village of Montaigut-le-Blanc, located close to the road between St-Nectaire and Champeix. Many of the buildings in these old towns date back to at least the eleventh or twelfth centuries.

still to be explored (Fig. 1.3). Artists and photographers are unlikely to run out of suitable subjects. The same is true of the geologist. Auvergne has always been a battleground of fire, ice and wind against the rocks, an area where humankind has ceaselessly fought against nature. Earning a living here has never been easy. Not surprisingly, the Auvergnats are craggy and durable, and have always been considered by other French people as mean and cagey. One writer has indelicately observed that the French view of Auvergne's farmers makes the traditionally caricatured Scot seem like a lunatic spendthrift. However, the Auvergnats have much in common with the Highland Scots, for both have had to eke out a living from a land that often yields desperately little for much hard labour. The little Musée de Vie Rurale in Chastreix offers a true taste of Auvergnat life – a highly enlightening experience.

In the heart of Auvergne is the Parc des Volcans (Volcanoes Park), arguably France's largest and most important regional nature park. Measuring 120 km from north to south and with an area of about 393 000 ha, its common denominator is, of course, volcanoes. Fewer than 100 000 people live within it, but those who do are passionate about protecting their region and ensuring that it remains vital and attractive. In the north, 80 or more recent volcanoes or "puys" dominate the landscape around Puy de Dôme, which rises to a height of 1464 m (Fig. 1.4: 1). In the centre of the park lies Puy du Sancy, an older volcano some 3 million years old and also Auvergne's highest point at 1885 m (Fig. 1.4: 2). Farther south again are the Monts du Cantal, moulded from a single massive volcano, the oldest part of which goes back some 20 million years (Fig. 1.4: 3). Here the loftiest summit (Plomb du Cantal) rises to 1855 m. Fingering out from the area of Puy Mary are 12 valleys in which much of the geology and certainly some of Auvergne's most stunning scenery are to be found. Linking the Monts-Dore and Cantal massifs is the volcanic tableland of Cezallier (Fig. 1.4: 4). Here the highest point, Signal du Luguet (1555 m) sits within a tableland of wide open spaces. Lying to the west is the old granite plateau of the Artense, between the Dordogne and Rhue gorges, and to the east lies the broad downfaulted valley of the River Allier, floored with relatively young sedimentary rocks, onto which a few recent volcanic flows and cones have encroached. Farther east is the wild forested plateau of the Forez. Also to the east lies one volcanic outpost, Puy-en-Velay, included here because of its spectacular setting and position on one of the main tourist routes through France (Fig. 1.4: 5). Located in Haute-Loire, this superb town, overlooked by a huge statue and a church on each of two spectacular volcanic spires, punctuates the lower ground on the eastern flanks of the Massif Central. This finds a place within the guide as its volcanism is linked with that of the main part of the Parc des Volcans.

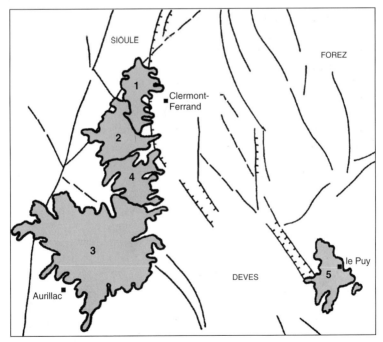

Figure 1.4 Extent of volcanic provinces in Auvergne. 1. Les Puys de Dôme; 2. Massif du Monts-Dore; 3. Cezallier; 4. Cantal; 5. Puy-en-Velay. Major faults are shown as solid lines with pecks on the downthrow sides of main graben.

7

Chapter 2

Preparing to visit Auvergne

Visiting any region for the first time is always exciting, but it can also be confusing and frustrating. This is especially true if one is going for a specific purpose, yet with a limited period of time to achieve one's ends. Having visited Auvergne many times, sometimes with family, sometimes with parties of geologists and sometimes as a walker or photographer – and always with an interest in local food and wine – I feel reasonably well equipped to proffer some basic advice.

Auvergne is not a region through which to rush. The topography prevents this and to drive through with little time spent in the fresh mountain air is to miss the main point of exploring this lovely region. At best, the pace of life is sluggish, so relax, spend time just experiencing its culture, its way of life and the feel of its wonderful open spaces, as well as investigating the geology. Then again, leave time to visit local markets, explore some of the old churches or castles, take in a museum or two, and certainly tarry awhile in local auberges and restaurants, sampling the local wines and the good-value plats du jour offered to the traveller.

Climate

The climate of the Auvergne is idiosyncratic, to say the least, and conditions can be very changeable. Thus, some summers are characterized by weeks of almost unabated warm sunshine, whereas during others the rain seems to fall every day. Generally speaking, there are fewer rainy days than in most parts of Britain, but when it does rain the quantities can be large, particularly during summer storms. It is common to read that Auvergne stays green all summer, being watered by storms and cooled by its considerable elevation; however, although this may have been true of September 1998, for instance, in the late 1980s there was hardly any rain for three summers in a row and the region turned quite brown.

In a typical year Auvergne receives plenty of warm sunshine, particularly towards the end of the summer and beginning of autumn. Spring is

9

usually brief, lasting typically for the month of May. At this time, while snow lingers in hollows within the landscape, only a few paces away daffodils, violets, orchids and other spring flowers burgeon amid green meadows grazed by well fed mountain cattle. Meadows of wild flowers flourish at this time and, if you are a keen botanist, this is a good time to visit. Autumn is perhaps the best time to visit, for it is often calm, warm and usually sunny. As the leaves within the wooded areas begin to turn, so the beauty of the region is enhanced. Mushrooms and other fungi emerge among the woodlands, and locals will be seen collecting wild mushrooms while the morning dew is still on the fields. Wild raspberries, strawberries and blackberries are also abundant at this time. Furthermore, the main summer tourist season is then over and hoteliers and their staff have time to give full attention to providing what they do so well – a range of typically Auvergnat cuisine, such as la truffade, coq au vin, tripou d'Auvergne and pounti, accompanied by a selection of local wines, such as Côtes d'Auvergne and St-Pourçain. During the winter months, the region is busy coping with winter sports enthusiasts and, from the geologist's point of view, it is not a good time to visit. In any case, many of the roads will be difficult of passage, with the higher outcrops covered in snow. With this in mind, it is best to be prepared for almost anything. Certainly take light clothing at any time between spring and late autumn, but also take warm sweaters, windproofs and cagoules. Indeed, I would recommend carrying warm clothing and wet-weather gear. Even in summer, winds can be fierce on the mountain tops, or mists can roll in, forcing the temperature down quite quickly. Footwear is also important. On some of the excursions the geology can be explored without recourse to anything other than stout shoes or trainers. However, at other times, rough terrain demands ankle support, so that boots become necessary. Special requirements for specific itineraries are described in the appropriate chapters.

As regards other leisure activities: if you enjoy a cooling swim, pack swimwear and towels, for several of the lakes are perfect for swimming, although the water can be cool in spring. It is also possible to swim in some rivers, but take care of warnings given where hydro-electric schemes operate; from time to time water may quite suddenly be released into some river systems, causing the rate of flow and height of water to increase extremely rapidly. For entering lava tunnels, take a hard hat and good torch; indeed, packing some protective headgear is not a bad idea when visiting some quarries too.

Transport

Auvergne can be reached from Britain in several ways. For instance, Air France offers daily flights from various UK airports to Paris (Charles de Gaulle) and thence Air Inter provides connecting scheduled flights to Clermont-Ferrand. Seats vary in cost between £220 and £260 per person return (2001 prices), but need to be booked well in advance as they are much used by the business community. It is then essential to hire a car. This is simple, since several major hire companies have an office at Aulnat airport (Clermont-Ferrand) and one can pre-book a vehicle from the UK. Arriving from outside Europe, the arrival destination again has to be Paris, for it is from here that most internal French flights depart.

French Rail can package train travel from London Victoria to Clermont-Ferrand or other major stations in France, a method of transport that takes much longer and necessitates changing stations in Paris. However, if you are a train buff, here is an excuse to travel through some interesting countryside and through some typical French towns in relative comfort. Taking a car via this route can be rather expensive, so hiring in France is probably a better deal. Trains also arrive in Clermont-Ferrand from destinations outside France; thus, visitors from Germany, the Netherlands, Belgium and Luxembourg may find they can travel directly there without suffering the congestion of Paris. This cuts down the travel time considerably.

If driving is preferred, use one of the ferry operators to cross the Channel (or Eurotunnel). If time is of the essence and money is no problem, then a short ferry crossing or Eurotunnel, followed by a drive down the toll motorway system (E 15–19 south to A 71–E 11) brings you to Clermont-Ferrand in less than a day. My preference is to do things more slowly, for instance taking the Portsmouth–Caen ferry, then cruising gently down the excellent N (nationale) or D (département) roads, taking a route such as Alençon–le Mans–Tours–Châteauroux–Montluçon. Then again, depending upon which part of the region you wish to visit, to reach farther south and into the Cantal, it is better to turn south at Châteauroux, crossing the lovely countryside via the D roads through Guéret, Aubusson, Ussel and Mauriac, by which route one can travel south quite quickly to Aurillac. This more leisurely method takes longer, but in a couple of days one can be in Auvergne – even the Cantal – and still feel fresh, with the bonus of having passed through some superb French towns and scenery.

Visitors from other European countries will find that a system of fast roads leads to Clermont-Ferrand. From Spain, for instance, the preferred route would be via Biarritz to Bordeaux along the E 5–E 70, and then east across country along the E 70–N 89. From Belgium and the Netherlands the best route is that via Lille and Paris. From Germany it is probably best to travel via Luxembourg and then join the autoroute through Nancy and Dijon. Swiss visitors from the south would probably need to get there via Geneva and Lyon, or, if resident in the Basel region, drive through Besançon to Dijon and then cut across country via Autun and Vichy.

Accommodation

There is no shortage of good hotels within the region. In the larger cities, such as Clermont-Ferrand and Aurillac, there is a wide range, from 4-star establishments to cheaper hotels or pensions. Luxury hotels are less common in smaller towns, but there are some excellent small hotels in the 2- and 3-star categories with fine restaurants. Good centres for the northern region are Gannat and Ebreuil (for the Gorges de la Sioule), and Royat and Orcines are ideal bases for exploring Puy-de-Dôme. Farther south and into the Massif du Monts-Dore, le Mont-Dore and la Bourboule offer a wide range of hotels and facilities, and in a more rural setting there are Chambon-sur-Lac, Murol and Besse-en-Chandesse. Within the Cantal – a far larger area – you should probably avoid Aurillac, since it is heavy with traffic these days, and go for smaller towns such as Thiézac, Vic-sur-Cère, Murat or St-Flour, depending upon exactly which part you want to visit.

An alternative to hotel living is to spend time in the excellent chambres d'hôtes (bed & breakfast). Many offer high-quality accommodation with a hearty breakfast in lovely locations, and some provide evening meals to order (these tend to be called chambres et tables d'hôtes). Several I have visited also provide a well equipped kitchen. Using these places, two people can live for about FF180 per day, including breakfast. It is also a good way to improve your French.

Then, of course, you can camp. There is a wide range of sites all over Auvergne, but it should be remembered than many open only in May and close on 15 September. If camping appeals, then it is a good way to experience Auvergne and it is certainly much cheaper than hotel living. The green Michelin *Camping caravanning France* is published every year and contains all you need to know about this style of living. Depending upon the star rating of the site (which is related to the range of facilities it provides), daily costs for a car, tent and two people range between FF65 and FF100. Some sites also offer huts or rooms at very reasonable rates and have their own bars and small restaurants. Bear in mind that concessions are often made in May and September, out of the peak season.

In Appendix 2 is a list of hotels and campsites that are reliable, comfortable and ideally located for exploring the geology and scenery. There may be many other excellent spots, but one cannot visit them all.

Chapter 3

The geology of the Auvergne

The sequence of events

The geological evolution of the Auvergne and the wider region we know as the Massif Central took place in three stages. About 600 million years ago, the part of the Earth's crust that now is Europe lay under an extensive ocean and was located at a quite different latitude from the present. On either side of this ocean lay two continents. Between 450 and 420 million years ago a **subduction** system was set up by changing convection patterns within the Earth's **mantle**, with the result that a slab of the ocean floor was inexorably consumed back inside the Earth, causing the ocean to shrink in area. This continued until the two continents collided between 400 and 350 million years ago, whereupon the marginal sediments were squeezed as in a giant vice, folded and uplifted. Accompanying this was strong **regional metamorphism** that generated massive volumes of **gneiss** and **schist**. The upshot was a new range of mountains injected with **granite**, which, in the hot and humid climate then prevailing, gradually became clothed in luxuriant vegetation. Over the next 100 million years or so, these mountains were worn down into an extensive undulating plateau in which rotting vegetation gradually accumulated in basins and gave rise to coal deposits. This was the end of the first stage, taking us to about 250 million years before the present time.

Despite difficulties in reconstructing events in Europe in the Late Palaeozoic, it seems clear that an ocean separated Europe from Africa during the Devonian period, which began about 380 million years ago. However, by the beginning of Carboniferous times, 40 million years later, this began to close as the two plates bearing Africa and Europe converged. At this time the southern margin of the European continent was bordered by sedimentary basins, and similar ones existed to the east, where an ocean separated Europe from Siberia along a line that became the great Urals mountain chain. The basins received huge amounts of sediment, initially eroded from the ancient Caledonian mountains that lay to the north. In time, the eroded debris was compacted into shale and sandstone, with lesser amounts of limestone and volcanic rocks. Indeed, in shallow nearshore regions, coral reefs flourished. As time progressed, Africa approached Europe, a process driven by movements in the underlying mantle, with the result that the basin deposits were squeezed together and dragged down along the line of a **subduction zone** similar in many ways to that currently operating along the western margin of the Pacific Ocean. What is not entirely clear is whether the subduction system saw Europe plunging beneath Africa, or vice versa; the evidence, such as it is, seems to favour the latter. Either way, the end result was the burial, deformation, recrystallization and re-emergence of these ancient sediments as the Hercynian mountain chain.

The mountain-building cycle, or **orogeny**, occupied the early part of the Carboniferous period, whereupon the basin-filling deposits were buried, intensely folded and metamorphosed. The chain of fold mountains – the Hercynian mountains – to which they gave rise was also intruded by huge bodies of granite, apparently in two phases, the first 350–330 million years ago, and the second 290–280 million years ago. It is these igneous and metamorphic rocks of Upper Palaeozoic age that outcrop in a broad curving belt largely located to the south and east of the region of Auvergne, but entering it in southerly reaches of the Cantal; they comprise what are termed the "schistes périphériques".

The basement below the main part of Auvergne, however, consists predominantly of older rocks that show the imprint of an earlier metamorphism dated at around 600 million years ago, during the later part of Precambrian time. This was allied to an older cycle of plate convergence, mountain building and metamorphism, and it is these rocks that generally form the basement upon which the Tertiary and Quaternary rocks of Auvergne were laid down or into which magma was intruded. This ancient basement material belongs to what is known to structural geologists as the "lémico-arverne zone". Its rocks show the effects of progressive metamorphism, that is to say, of recrystallization and deformation related to increasingly great depth of burial within an **orogenic belt**.

At the close of this first stage, because the land had been worn down, Europe was very widely invaded by ocean. Certainly, during the Middle Jurassic (about 180 million years ago), the Cantal area was beneath the sea; we know this from the rocks of this period and the fossil remains they contain. It then gradually emerged as further upheavals affected the crust, the seafloor sediments were eroded and lagoons developed. In these, marine animals thrived, then died, became fossilized and left their remains. Eventually, about 35 million years ago, the lagoons dried up and extensive **evaporite** deposits accrued, producing gypsum beds (to be seen in the Cantal region).

About 30 million years ago, a further spate of violent upheaval threw up the fold mountain chains of the Alps and Pyrénées, something that had a severe effect upon Auvergne. During this episode the ancient granite plateau became cracked, tilted and faulted. In the lower-lying areas of the Limagne and Ambert, sedimentary rocks were laid down. Similar rocks are to be found in the valley of the Dore and are scattered elsewhere in smaller depressions. Where the plateau had been cracked and fractured, molten magma invaded it, rising to the surface in pulses. Early eruptions generated volcanic terrain within the area of Limagne 20–12 million years ago, then built volcanoes in Velay and the Cantal 11–2 million years ago, and around Sancy 2.3–0.2 million years ago. Volcanism also occurred in Cézallier 6–3 million years ago. The peaks of the Cantal are the remnants of one massive **stratovolcano** that was the largest in France, measuring 100 km across and rising some 3000 m above the plains. The Sancy–Monts-Dore volcano was similar but slightly smaller.

The third stage began a mere 3 million years ago, when the Chaîne des Puys of the Clermont-Ferrand region grew up. This chain of 112 volcanoes has classic volcano morphology. The Puy-de-Dôme volcanic chain contrasts nicely with the older eroded volcanoes of the other two areas, and some of its more extensive flows are said to be rivalled in extent only by those found in Alaska and New Zealand. The most recent eruptions occurred as little as 7000 years ago. A simplified map of the geology is shown in Figure 3.1.

The rock types found within the Auvergne include **basalts, phonolites, andesites** and **trachytes**, many of which make beautiful specimens, containing large **phenocrysts** of augite, sanidine or hornblende. The **volcaniclastic** rocks include debris flows, **pumice** sheets, **pyroclastic flows** (**ignimbrites**), ash and surge deposits. Volcanic bombs and cinders abound, whereas the more explosive eruptions have brought up deep-seated rocks such as **peridotites** and **charnockites**. The episodes of mountain building generated extensive areas of granite, granitic gneisses and **migmatites**, and a range of micaceous schists. There are sedimentary rocks too, some containing abundant fossils. These tend to be found in the marginal regions, but are also found within the volcanic outcrop, for instance in the Cantal and near le Mont-Dore.

The basement rocks and their metamorphism

Those rocks exposed in the region of the Sioule and in the southern Cantal that have been affected by the older metamorphic event belong mainly to the **amphibolite facies**. That is to say, they are of moderate metamorphic

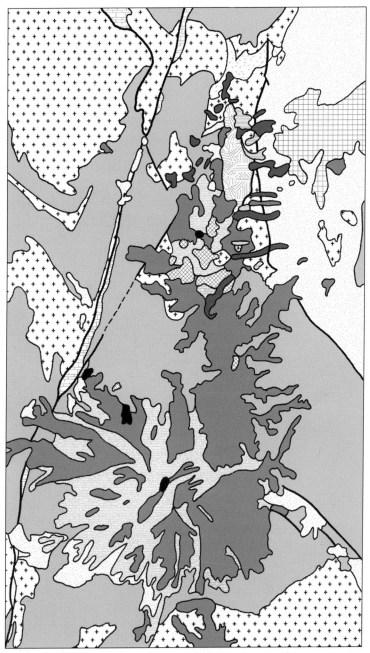

Figure 3.1 Simplified geological map of Auvergne. Thicker lines represent major faults. See opposite page for the key.

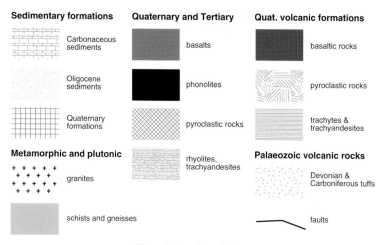

Figure 3.1 continued   Key.

grade, having been raised to moderate temperatures and pressures while being buried in an emerging mountain chain. However, some rocks have been converted into **greenschists** – a lower-grade assemblage – and others to the **granulite facies,** which are of much higher grade. These represent rocks that were buried respectively less deeply and at greater depth in the sedimentary pile. The fact that sometimes the lower-grade rocks lie structurally below the amphibolite-grade rocks or even the granulites poses a problem of interpretation. One explanation is that the inversion of the sequence is attributable to intense folding and dislocation, a process that could bring the more profoundly changed rocks above those formed at lesser depth, especially if **recumbent folds,** or even **nappes,** were formed. Indeed, this seems the simplest way to explain things; however, detailed work suggests it may not be as simple as this, and structural geologists continue to argue about what produced the observed field relationships.

The rocks of lowest grade are schists containing chlorite and sericite; these, in turn, pass into somewhat higher-grade mica schists containing two micas (biotite and muscovite). Next, and formed at greater depth, are two types of gneiss: gneisses with two micas (formed at moderate depth) and gneisses with biotite and sillimanite (a higher-grade assemblage of minerals presumed to have been metamorphosed at greater depth where temperature and pressure were even more elevated). Finally, there are migmatites, which represent rocks on the verge of melting and becoming remobilized by heat and pressure; these come in two varieties: those with the mineral assemblage (biotite and sillimanite) and those containing biotite and cordierite. This rock group can be studied during Excursion 1 in Chapter 5 (pp. 32–36).

Those rocks that also have been affected by the younger Hercynian orogeny show the effects of **retrogressive** metamorphism; that is, the minerals formed during the earlier high-pressure–high-temperature event, show the effects of recrystallization to other phases stable at lower temperature and pressure. Thus, early-formed biotite mica breaks down to chlorite, sillimanite to andalusite or cordierite, and so on. Many of the basement rocks of the region show the effects of this younger orogenic event, which reached quite high temperatures but substantially lower pressures than the earlier one.

The granitic rocks of Auvergne

Most of the granites found within the region were formed during the Hercynian events. There are four principal types:
- intrusive leucogranites containing two micas (biotite and muscovite)
- migmatitic leucogranites with two micas
- intrusive **monzonitic** granites and granodiorites (more plagioclase than potassium feldspar)
- anatectic monzonitic granites and granodiorites.

Within Auvergne the predominant types are the monzonitic granites and granodiorites. Such rocks rose either as diapirs into the crust and crystallized slowly, generating coarse-grain rocks with granular and often porphyritic textures, or they were generated by **anatectic** processes (i.e. remobilization deep within the orogenic belt) and, as a result, may have developed weakly foliated fabrics in response to crystallization or mobilization under high pressure.

Sedimentary rocks

Although Carboniferous rocks are widespread in some parts of France, Auvergne has relatively few deposits of this age. Restricted coal deposits occur in local basins, and there are also a few anthracitic tuffs. Those of Decazeville, in the south of the Cantal, are visited in one of the itineraries described in the Cantal section of this guide.

Deposition of Oligocene rocks in Auvergne was strongly asymmetric, with most sedimentation occurring in marginal troughs and basins, particularly in the east. In the downfaulted region of Limagne, substantial areas of calcareous sediments and marls are to be found. Smaller basins occurred in the Cantal, particularly near to Arpajon, where shales and calcareous rocks accumulated. Oligocene basins are also found near

St-Flour. Elsewhere, for instance in the vicinity of le Mont-Dore, lake deposition occurred over much more restricted areas. There are other isolated basins in which sediments are to be found; for instance, Miocene lake beds are found near St-Gérand-le-Puy, and marls and sands can be found in fossil valleys near Gergovie and near le Puy-en-Velay.

The Lower Oligocene strata are shallow marine deposits, some of which were lagoonal in character and may include evaporites. Such lagoonal deposits can be inspected near Aurillac. Upper Oligocene beds, on the other hand, represent a more widespread inundation of parts of the region, particularly the Grande Limagne, which remained a downfaulted trough throughout the period and into which extended an arm of the sea. This represents the largest area of continuous sedimentation during the period, a fact borne out by the presence of a monotonous sequence of Cypris-bearing **marls** that is at least 1 km thick (Cypris: a kind of crustacean called an ostracod, found only in rocks younger than Oligocene).

During the Quaternary, ice sheets spread over the higher parts of the region and, on their retreat, left a mixture of **erratics**, **moraines** and melt-water deposits over large areas. Such an assemblage of rocks, together with the erosional legacy of glaciation, particularly U-shape valleys, are widely observable within the Cantal and Sancy–Monts-Dore massifs.

Recent volcanic rocks

The recent volcanism within the Massif Central is dominated by some-what alkalic rocks (e.g. basanites and alkali basalts; calc-alkaline basalts predominate in the orogenic or Alpine tracts). Initially these rocks were expelled from fissures in the granitic basement. In time, however, more differentiated magmas evolved, giving rise to a suite of rocks ranging in composition from **hawaiite** to **rhyolite**. These appeared when activity became centralized and were the result predominantly of fractional crys-tallization rather than of crustal contamination. In both Monts-Dore and Cantal, petrographic evidence strongly supports the argument that basic magma frequently entered magma chambers that were already differen-tiated, with the result that partly crystallized enclaves of basaltic matter are to be found encased in either trachytic or phonolitic material. Further-more, there is every reason to believe that the intermediate (andesitic) types found within the region are the result of homogenization of silicic and basic fractions. Modern research interprets the chronological, petro-logical and structural characteristics of Auvergne as the result of the gradual ascent of a mantle **diapir** through a crust affected by the Alpine fracturing that took place in Oligocene–Miocene times.

Chapter 4

The itineraries and their scope

The excursions have been arranged by geographical area, beginning in the Sioule and Puy-de-Dôme in the north, moving through the Sancy massif in the centre and thence to the south, to the Cantal, Cézallier and le Puy. It has been arranged this way since most travellers from Britain and northern Europe will approach via Clermont-Ferrand, but there is no real reason to study the geology in this order. Furthermore, you can mix and match excursions in any way you please. Whereas some excursions will focus on a particular style of structure or topic, others will reveal a range of rock types and geology exposed in travelling along one of the obvious cross-country routes. The amount of geological detail also varies. In some excursions the input is fairly intense; in others there is as much scenic, cultural or botanic input as there is geology. I do not apologize for this, but I do make it clear in the description that this is so.

As an aid to finding your way around the region, I can recommend a variety of excellent French maps, available in France but also obtainable in Britain and elsewhere, to order. These are published by the Institut Géographique National (IGN) and are inexpensive, of high quality, reliable and up to date. The following will be found particularly useful:

- For an overview of Auvergne, *TOP250* (scale 1:250000 or 1 cm:2.5 km) sheet *111* (Auvergne) is excellent.
- To plan routes in more detail, *Série Verte* 1:100000 (1 cm:1 km) sheets *42* (Clermont-Ferrand–Montluçon), *49* (Clermont-Ferrand–Aurillac) and *50* (St-Etienne–le Puy) are good.
- For use with the excursions, three other IGN maps are vital. These are all in the *TOP25* series on a scale of 1:25000 (1 cm:250 m): *2531 ET* (Chaîne des Puys), *2432 ET* (Massif du Sancy) and *2435 OT* (Monts du Cantal).

Some itineraries are arranged so as to provide circular excursions; others start at A and end at B, providing a choice either of reaching a new base or returning to the start point. Again, since a full exploration of Auvergnat geology means crossing from one volcanic zone to another, there are some itineraries that provide useful links between Puy-de-Dôme and either Sancy or Cantal, and so on. Knowing that people will explore at differing rates and in different ways, I have given some indication of how long each tour may take to complete. This is likely to be more reliable a guide where there is relatively little walking involved, for, once travel is largely by foot, the pace can vary widely between one person and the next. As a regular mountain walker who carries little body weight, I suppose my timings will be quicker than some, but I think they will give a reasonable guide to most people. In any case, where walks follow parts of the French footpath system, posts often suggest times required and, generally, these assume a fairly steady but leisurely pace.

I have purposely included some excursions that have to be completed entirely on foot. Auvergne is magnificent walking country and, weather permitting, you should certainly try to include one or more of these. Some are circular, others necessitate transport to be provided at either end (appropriate if you are in a party using a coach, less easy if you have just one car). In some cases a tour can be completed in more than one way; thus, while on Sancy summit, it is possible to return to base via cable car and coach or car, or take a two-hour walk down to a col and then return to base. I have tried to make it clear in the appropriate sections what the full options are.

It will be obvious that more than one of the shorter excursions can be linked to give a varied and full day's geology with another. In places I have made suggestions as to which make good partners. In some cases an excursion will pass by or end up at a town, historic building or museum that can provide variety over and above geology (always a healthy plan). Since I have always enjoyed a mix of leisure and geology, as well as time for photography, you may find within an itinerary some time allowed for a cooling swim, or a lunchtime meal at a local hostelry. Clearly, these are simply suggestions, and are not a vital part of any masterplan. Each individual will doubtless want to edit my suggestions to fit his or her predilection, the weather conditions, or the time available.

Please respect the wishes and habits of the local inhabitants; close gates, and do not climb over walls, chase cattle or enter quarries with large DEFENSE D'ENTRER signs (unless you can find a workman or owner whom you can ask for permission to enter). Sometimes it may not be possible to park a vehicle safely near to the start of a walk or immediately outside a quarry entrance. Please think carefully about where you park, as many Auvergnat roads are narrow and twisting, and bad parking may contribute to a nasty accident. Where appropriate I mention good parking spots and other laybys large enough for a car or coach.

One thing I must also mention, as it can lead to extreme frustration, is to remember that this is France and that most shops and many museums, visitor centres and castles close between midday and either 14.00h or

16.00h, and may also be closed on Mondays or Tuesdays. Furthermore, some places, particularly restaurants, may not open on Mondays. It is always wise to check opening times before planning to visit monuments or museums, even though I have given some information within these pages and some useful addresses and phone numbers in the appendices. Also remember that opening times may vary with the time of year.

Finally, bearing in mind that at least some things are not forever, and that books are written and do not reach the shelves of booksellers in an instant, although the details given were correct at the time of writing, things may have changed. For instance, quarries once readily accessible may become closed to the public; a new road cutting may appear, or an old one suddenly have been manicured into a picnic spot with toilets and swings. I hope that everywhere mentioned will still be available for inspection, but I cannot guarantee it. Should you find changes have occurred, or improvements made to access, then please be sympathetic; indeed let the publisher know, and we can amend future editions of this book. In this way the guide can be kept as accurate as is humanly possible.

Chapter 5

Puy-de-Dôme and Allier

The general structure of northern Auvergne is that of an extensively faulted granitic massif – the Plateau des Dômes – with younger sedimentary rocks on either side. The latter occupy downfaulted regions flanking the crystalline basement **horst** (Fig. 5.1). The downfaulted valley to the east – the Limagne – defines the course of the River Allier, and to the west, the River Sioule occupies a similar structural trough. The word "limagne", in fact, is a common noun that denotes a low-lying fertile area. The limagne adjacent to Clermont-Ferrand is properly known as the Grande Limagne, to distinguish it from those of Brioude and Issoire. Covering much of the granitic massif are a variety of volcanic rocks of post-Pliocene to Recent age that give rise to the famous volcanic puys and the extensive cheires or lava flows. It is these that dominate the landscape and which provide the main focus for the itineraries in this northern region of the Massif Central.

The start of the latest phase of volcanic activity in this area was marked by **hydromagmatic** eruptions of Aquitanian age – beginning about 24.5 million years ago – that punctured the sedimentary floor of the Limagne. Argon dates for basaltic rocks at Gergovie, the Côtes de Clermont and at Châteaugay yield ages of around 16 million years ago. Basaltic flows in the Montagne de la Serre, and at Charade give ages of between 3.5 and

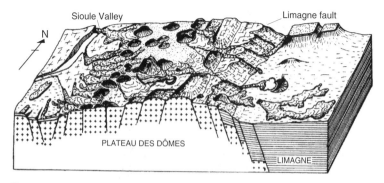

Figure 5.1   Cross section of Puy-de-Dôme from west to east, showing the faulted granitic horst and the flanking valleys of the Sioule and Allier (© de Goër de Hervé & Camus 1991).

25

3.0 million years. The oldest flows of the Chaîne des Puys proper were extruded 70 000 years ago. The effectiveness of radiometric dating techniques diminishes with younger rocks, so new methods of dating, including **thermoluminescence**, have enabled geologists to establish ages for such features as the Maar de Clermont (156 000 ± 17 000 years) and rocks in the Ardèche (to the southeast), where ages of 80 000 years have been established; similar techniques have now been applied to the younger flows and domes of the Chaîne des Puys and we now have a pretty good idea of their chronology. Since those early eruptions in the Limagne, over a hundred volcanoes have been active in the region, the most recent – giving rise to the crater in which Lac Pavin now sits – having occurred about 6000 years ago.

Granitic and gneissose rocks underlie the valley of the Allier, but are downfaulted and covered by more recent sedimentary rocks. The line of faulting strikes roughly north–south and is marked by a distinct scarp with a slope of about 30° and a relief of hundreds of metres. The floor of the Limagne fault trough lies at an altitude of 300 m above datum, and the highest part of the plateau rises to 1030 m. The easterly bounding fault of the plateau is steeper and more marked than its counterpart to the west, a feature dictated by the structure of the basement because it is much less faulted. The summit of the plateau lies about 4 km west of the bounding fault, and the broadly aligned volcanic summits follow the same trend but at 2–3 km to the west of the axis.

Although the floor of the Allier Valley is low and flat, it is not featureless; by standing on the granite scarp in suitable locations, it is possible for the observer to see that some flows and some volcanic necks have either covered or punctured it. Thus, prior to eruption along the main volcanic chain there were sporadic episodes that gave rise to a variety of volcanic rocks that have been thrown into relief by differential erosion. One can pick out dykes and neck-like structures at Montrognon, Puy Giroux and Montaudoux (near Clermont-Ferrand), and in the Limagne the weathered remnants of old necks of pépérite (basaltic breccia mixed with calcareous mud) are to be found, such as at Verthaison. Then again, the Miocene–Pliocene basic flows of Gergovie and the Montagne de la Serre, once occupying valley floors, now stand proud as inverted relief, after the enclosing softer rocks have been stripped away.

The overall trend of the Chaîne des Puys is north–south, but secondary fractures with NNE–SSW trends play a part in dictating where the volcanoes have developed. Naturally, the path and form of lava flows that emanated from the fractured basement were much dictated by the existing form and slope of the pre-volcanic surface. To the west, away from Limagne, the flows tend to spread out laterally over wide areas, and to the east they tended either to flow between existing cones or to enter steep-sided valleys incised between the plateau summit and the Limagne.

The study of samples from the basement area reveals the range of rock types present. There are schists and ancient volcanic rocks of Devonian to Early Carboniferous age (c. 360 million years); then there are granitoid intrusive rocks of Hercynian age (Permo-Carboniferous), and gneisses and migmatites that pre-date the Devonian period. In addition, exotic blocks brought to the surface by violent eruptions (e.g. Puy de Beaunit and Puy de Chanat, also at Puy de la Nugère and Tunisset), include charnockites and granulites from the base of the granitic crust, and peridotites that have come from the upper mantle (Fig. 5.2).

The lavas of the Chaîne des Puys form a more-or-less continuous petrographic series, although this is not always apparent in the field, where one can apparently distinguish two principal groups. The more **melanocratic** (grey or black) types comprise a group of predominantly basic to intermediate rocks relatively poor in olivine (43–55% SiO_2) that include basanites, basalts, leucobasalts (**hawaiites**) and trachyandesites (**mugearites**). The second group is lighter in colour (more **leucocratic**), tending to be light grey or whitish, and is represented by trachytes and rarer **rhyolites**. These have silica contents greater than 60 per cent. That they were available for eruption at much the same time – and therefore form part of a continuous petrographic series – is made evident by what can only be described as "mixed magma" rocks, which can be found at some localities (e.g. Puy de la Nugère).

Figure 5.2 Granite and charnockite blocks within explosive volcanic deposits at Maar de Beaunit. Excursion 3, stop 10 (pp. 44–44).

27

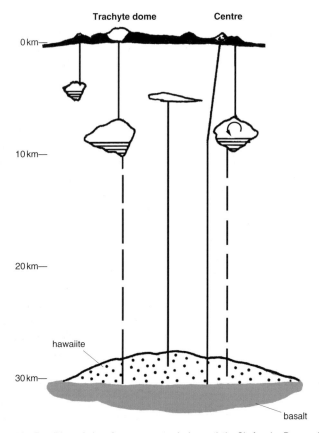

Figure 5.3 Possible evolution of magma reservoirs beneath the Chaîne des Puys region. The trachyte dome could be Puy de Dôme, whereas the centre might represent Ancien or Nouveau Nugère. After Aubert & Camus (1974) and Villemont et al. (1980).

Modern research suggests that the magma that fed this huge chain of volcanoes did not come from a single chamber; rather, there were several reservoirs that occupied a position between Lac d'Aydat in the south and Puy de la Nugère in the north, and at a relatively shallow depth (estimates range from between 2 km and 6 km). There was probably a network of interconnected chambers dispersed within the granite carapace in this region, these being sourced by larger differentiated magma reservoirs at depths between 25 and 30 km (Fig. 5.3). The shallow chambers probably represented **neutral buoyancy zones**, where rising magma ponded into small reservoirs when the magma pressure was balanced by the hydrostatic pressure. If such high-level chambers existed for lengthy periods, clearly there would have been sufficient time for them to differentiate by crystal fractionation, prior to erupting at the surface. By this mechanism the same reservoir was able to source different volcanoes and styles of eruption, ranging from basaltic maars to trachytic cumulodomes. In some instances, deep-seated magma may have reached the surface without stalling at all on its way to the surface. Certainly there is field evidence for the fact that both silicic and basic magma was available at one and the same time at some eruptive centres (e.g. Puy de la Nugère).

Various seismic studies have been carried out in recent years and reveal that below the region of the Puys there is no trace of the **Mohorovičić discontinuity** (the true boundary between the Earth's crust and mantle); rather, at a depth of 23–29 km, the velocity of seismic waves averages 7.2 km/s, in a kind of transition zone between crust and mantle, perhaps representing a pocket of abnormal mantle. Deeper, between 29 and 44 km, the velocity of seismic waves increases to 7.2–8.4 km/s, indicative of the mantle material. Further geophysical evidence appears to show that there are two zones of low viscosity below the Chaîne des Puys; one lies at a depth of around 2 km, the other somewhat deeper, at 5–6 km, between the Gour de Tazenat and Puy de la Nugère. One interpretation of these is that they represent reservoirs of partly crystallized magma left over after the main volcanic episode. Interestingly, the modern region of highest heat-flow is recorded not within the Chaîne des Puys but near le Puy-en-Velay. Exactly why this is so is not clear, but there seems little reason for the inhabitants of that town to fear imminent volcanic activity.

The principal difference between the two types, and the one with the most profound effect on their mode of eruption, is their differing viscosity: the more silicic rocks are more viscous and generally richer in gas. Thus, they have a propensity for being explosively erupted, often as pyroclastic flows, or intruded as viscous domes. Where groundwater is involved, such magmas may erupt with very violently, giving rise to explosion craters or maars, which have low rims and flat floors that often fill with water to form volcanic lakes (e.g. Gour de Tazanet, Lac Pavin). The basic and intermediate types of lower viscosity may simply emerge as flows, but more often were erupted via Strombolian events, sending up clouds of ash, cinders (scoriae) and blocks that fell back around the vent, building steep-sided cones (about 80% of the puys in this region). Some of the more energetic explosions threw out large bombs (up to 1 m in diameter), which can be inspected at many accessible localities.

Most of the scoriae cones have (or had) craters at their summits and, if multiple eruption focused around the same conduit, may show more than one crater, perhaps one within the other, or one offset slightly from the other, indicative of a slight shift in the focus of the eruption. In travelling across the puys, one can see all different sorts and combinations of volcanic landforms, from single craters, through multiple craters, into cones that have grown inside maars or domes that have blown apart, rather like those at Mount St Helens and Soufrière (Montserrat), and domes that have breached their skins, leaking viscous lava from their sides.

Lavas that were erupted more gently, being extruded as coherent flows, took on a variety of forms. The less viscous flows (basalts and trachybasalts) tend to be thinner, wider and longer than their more silicic counterparts, which tend to spawn short, thick, stubby flows. Many of the basic flows travelled along tubes, wherein molten lava flowed beneath a carapace of solid chilled lava, this acting as an efficient insulator and allowing the magma to flow more efficiently than if it simply moved across the ground in contact with the atmosphere. Most Auvergnat basic flows have blocky or **aa** surfaces (cheires) and are often strongly vesicular. Many flows – and this applies to both groups – developed columnar cooling joints that look like serried ranks of huge organ pipes, similar to the Giant's Causeway in the north of Ireland. Particularly famous ones occur at Bort-les-Orgues, St-Flour and at Roche Sanadoire.

So much for a brief history of the geology of the region covered by the first group of excursions. For those interested in the chemistry of rocks, analyses of three typical rocks from the Chaîne des Puys are given in Appendix 4. Now it is time to explore it on the ground and see exactly what did go on and how this affected the Auvergnat landscape over the northern area. The following itineraries endeavour to provide the visitor with as wide a range of geological experiences as is possible in this fascinating and visually exciting region.

This group of excursions is located in the northern Auvergne, between the Valley of the Sioule in the north, and Lac Chambon, in the south (Fig. 5.4). The lower part of the Sioule's course slopes very gently, with several islands set within the stream. The upper course is steeper, in places flowing quickly through deep gorges that were eroded as the marly Oligocene basin was lowered, thereby incising the harder granite basement beneath. The Ebreuil–Châteauneuf-les-Bains section is particularly scenic. It is this section that is followed in the first excursion described below.

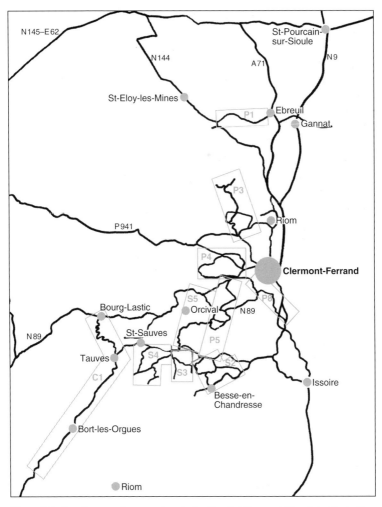

Figure 5.4 Location map for excursions in the Puy-de-Dôme and Massif du Monts-Dore regions. Prefixes: P, Puy-de-Dôme; S, Sancy; C, Cantal.

Excursion 1
Gorges de la Sioule

Introduction
This is an attractive half-day tour that explores a part of the valley of the Sioule where the river cuts a deep gorge through the resilient granite and metamorphic rocks of the Massif Central. It provides a good opportunity to study the basement that is now largely covered by volcanic rocks farther south. If you enter the Auvergne from the north, this excursion provides a lovely approach to Manzat, Gour de Tazenat and the northern part of the Chaîne des Puys.

Starting point
The starting point is the small town of Ebreuil, just to the east of the point where the river enters the gorge section. Leave the town by the D915, which follows the course of the river along its north bank (Fig. 5.5).

Itinerary
Proceeding westwards, the road passes through a broad valley with a granitic plateau capped with limestone to the north and the river on the

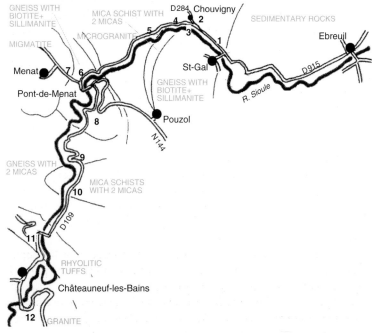

Figure 5.5 Locations and simplified geological sketch map for Excursion 1.

left-hand side. For the first 6–7 km, the river winds through a pleasant verdant valley, with occasional roadside outcrops of weathered granitic gneiss.

1. Approaching Chouvigny, the sides steepen and the river enters the gorge proper. Once it does so, there are several parking places that allow access to viewpoints, many of which are approached by steps to hand-railed platforms. These give splendid views of the many tall pillars of well jointed gneiss and granite, the steep-sided gorge walls and the fast-flowing river below (Fig. 5.6). The detached pillar seen as one enters the gorge section is known as Roc Armand and can be climbed by some steps. The basement rock itself varies in colour and texture, but typically is strongly porphyroblastic and banded.

2. Shortly, a steep slip road climbs up the north wall of the gorge, towards Château Chouvigny. Take this road (D 284) and just before the castle note a small quarry on the right-hand side. There are some good exposures here of **isoclinally** folded gneiss with both light and dark mica.

3. Return to the road, which runs along the river, and turn right towards Menat. Just through the houses of Chouvigny (basse), roadside outcrops of schist with two micas can be inspected and eventually the microgranite comes in, from beneath the schist outcrop. At the contact, **boudins** and **lenticles** of quartz-rich material are interspersed within the schist, and it appears that the foliation within the schist and flow fabric of the granite share a common dip, as if **concordant**. Since the microgranite has the form of a **laccolith**, it is likely that the relationships indicate differential movement between the two rocks adjacent to the margins of the laccolith.

4. Continue westwards along the D 915 to a quarry on the right-hand side. This is located about 300 m prior to the short road tunnel. This is one of the best exposures in the microgranite, which here is cut by a dolerite dyke. The rock is strongly porphyritic, with a light grey colour; the groundmass is granular. One of the most interesting features is the variety of enclaves within the rock. Rounded-to-elliptical igneous material is common and has a microdioritic composition. Then there are schistose enclaves that, when viewed under thin section, are revealed to contain cordierite, andalusite and sillimanite with or without corundum and spinel.

5. Close to Hôtel des Roches (which has a lovely restaurant on the opposite side of road, overlooking the Sioule) there are two red-painted houses. Park just beyond these and inspect the outcrops of strongly porphyritic grey granite in the roadcuts. The large white feldspar phenocrysts attain diameters of 2–3 cm and the rock is quite unweathered. Variations in granite texture and mineralogy are evident.

Figure 5.6   Gorge de la Sioule: jointed granite cliffs rising above the River Sioule.

6. On approaching Pont de Menat, stop just prior to the first houses, where outcrops of strongly banded gneiss containing biotite and sillimanite can be studied.
7. At the Pont de Menat, the D 915 crosses the main N 144 road. Here turn right prior to the bridge and drive for 500 m along the N 144. Having negotiated a major dog-leg bend in the road, an old disused quarry appears on the next bend. Take care in parking here, as the visibility is poor. The walls are made from a pinkish gneiss with an almost migmatitic appearance, which is indicative of mobilization during orogeny. The strong foliation is picked out well by the dark micas which alternate with K-feldspar and muscovite. It is very rich in both biotite and K-feldspar.
8. Return to the Pont de Menat and turn right towards Châteauneuf-les-Bains. Cross the river and rejoin the D 915; the fine old bridge over the river, now bypassed by the main road, can be seen off to the right-hand side at this point. The road now follows the east bank of the Sioule, winding tortuously above the meandering river in the valley below. In about 2 km, a large parking area lies on a sharp left-hand bend. Pull off and take the blue-marked footpath opposite, climbing up the valley sides to the ruins of thirteenth-century Château Rocher. This occupies a dominant position high above the gorge and offers spectacular views of two major meanders, with marked depositional/erosional regimes on the inner and outer banks respectively. Return to the road.
9. Continue along the road until, about 1.5 km before the village of Lisseuil (delightful Romanesque church with thirteenth-century statue of the Virgin), where the road runs along the south side of a large meander, roadside exposures of gneiss with two micas come in. Over the next few hundred metres, considerable variation in texture and mineralogy can be followed: sometimes the rock is coarse and feldspathic, elsewhere it is rich with greenish amphibole. The foliation dips constantly at 45° towards the north.
10. About 1 km farther along the same route, outcrops of schist with two micas appear.
11. Pass the turning off to Ayat-sur-Sioule (bridge over the river here) and enter a straightish stretch of road. A completely different rock type follows the road. This is the outcrop of rhyolitic tuffs of Visean (Lower Carboniferous) age. The best section is to be found along the narrow road that turns off left towards Blot l'Eglise. It is a dull-looking fine-grain tuff with small **clasts** of quartz and feldspar.
12. Châteauneuf-les-Bains is a sleepy thermal resort with a modern spa building and is pleasantly located beside the Sioule. Twenty-two springs are used here, providing both cold water for bottling and hot

water for the baths (temperatures reach 36°C). Just prior to reaching it, one can take a sideroad that leads to a large rotational landslip (signposted). This can also been seen from the town itself. This sideroad (D 227) is a direct way to reach Charbonnières-les-Vieilles and Gour-de-Tazenat. On entering the town, one of the more prominent landmarks is a small granite boss surmounted by an ornate cross. Climbing up the path here affords an excellent view of the large meander in the Sioule. To the west the walls are etched from the granitic basement that also forms the lower part of the gorge to the east; however, above this are steep cliffs composed of the rhyolitic tuffs. This is a spectacular point at which to end the excursion and take some refreshment.

Notes

At Ebreuil there is a fine campsite right beside the river, about 0.5 km out of town and beside the D 915 (Camping le Filature), which also has rooms for rent. There is the Hôtel du Commerce in Ebreuil itself, which, by the way, has a fine abbey church with a Romanesque nave and transept and early Gothic choir with radiating chapels. Château Chouvigny is also worth a visit if you like well restored castles (open Easter to 30 September, between 9 h and 12 h, and 14 h and 18 h; FF10 entry fee). It also is a fine viewpoint over the gorge below.

Excursion 2
Royat–Orcines walking tour

Introduction

This tour can be followed one way only or made into a good full-day excursion, starting either at Royat or Orcines–la Baraque and returning to the starting point. Here the excursion is described starting from la Baraque. It provides fine views over the Allier Valley, and of the town of Royat and the lava-capped granite plateau. It also enables study of basement granite, an **aplitic** intrusion, maar deposits and volcanic flows in the valley of the Tiretaine River.

The town of Royat is an elegant thermal spa and was known for its healing properties by both the Celtic Arverni tribe and the Romans. However, it was during the nineteenth century that they really became fashionable, with Empress Eugénie gracing the spa with a visit in 1862. Until the start of the First World War, it was popular with royalty. Five springs are tapped, the hottest water attaining a temperature of 31.5°C.

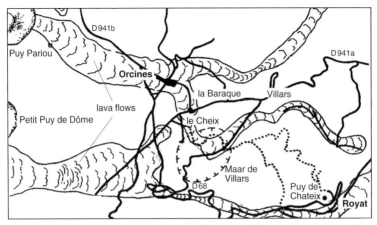

Figure 5.7 Sketch map of geology and routes for Excursion 2. Footpaths are shown by dotted lines.

Geology

The 8580-year-old trachyandesite lava that was erupted from Puy de Pariou first flowed northeastwards from the vent, then came against the granite plateau and was deflected towards the east; subsequently it was deflected again, this time by a small granite boss, upon which now sits the church and hamlet of Orcines (Fig. 5.7). It then advanced towards la Baraque (the main road is built on the surface of this flow) where it was deflected by another granite boss capped by basalt and scoriae, which, incidentally, was the source for a much older basalt flow that gave rise to the Plateau de Prudelles. The lava split around this knoll, one branch entering a narrow valley that runs towards Clermont-Ferrand, the other a steep-sided narrow valley in which the hamlet of Villars now sits. The latter branch of the lava almost filled its valley and eventually ceased to flow at a point called Fontmaure, where it forms a 15 m-high flow front.

A much older series of basalt flows was erupted from Petit Puy de Dôme (44 000 years ago) and swept eastwards, entering the steep-sided valley of the Tiretaine River in which the town of Royat now sits. These flows poured over the edge of the granite plateau and then spread out in two lobes, one towards Chamalières, the other through Beaumont towards Plateau des Cézeaux.

Starting point

Start at the Hôtel Relais-des-Puys, la Baraque, on the D 941a road. This point lies about 4 km west of Clermont-Ferrand.

Itinerary

From la Baraque, walk westwards for about 200 m, then turn left down a narrow metalled road towards the hamlet of le Cheix. The summit of Puy de Dôme can be seen looming up above the plateau.

1. Blocks of trachyandesite can be seen on the roadside just after turning off the main road. After passing the fountain and stone troughs, continue down hill past the houses and fork left onto a track through the trees (marked as a footpath with coloured paint and marked "Villars").

2. Shortly after joining it, note a small quarry on the left with banks of reddish cinders overlain by soil. This is a part of the rim of the Maar de Villars that punctuates the granite basement and that can be discerned at several points as the walk proceeds. Turn left along the road and in a short distance join a footpath meeting the road that comes in on the right-hand side. This path leads to Royat and at several points there are small outcrops of weathered porphyritic granite. Where there are clearings in the trees on the left-hand side, it is possible to see, on the opposite side of the valley, the granite-capping trachyandesite flow from Puy Pariou.

3. Ignore the first right fork, but at the second, turn right, southwards, signposted to le Colombier. The marked path then heads towards the granite edge and the steep-sided valley of the Tiretaine River. In about 20 minutes, note a Michelin sign pointing up the hill on the right-hand side towards Puy de Chateix. Make the short climb (beautiful orchids here in May) to the closely jointed outcrops of mica-bearing quartz–feldspar aplite. This appears to be a small granitoid boss. From the summit mast there are fine views down over Royat old town and its superb church, and towards Puy de Gravenoire with its huge cinder quarry. On the descent into Royat, just above the houses, sketchy outcrops of Oligocene arkoses occur.

4. Return to the path and continue down hill along the track. Eventually reach le Paradis and its posh hotel/restaurant. The views from here are very fine. Just opposite the hotel is an abandoned quarry in the weathered basement granite. Descend towards Royat.

5. Roadside outcrops on the right-hand side expose fresh basalt of the 44 000-year-old Petit Puy de Dôme flow that partly filled the Tiretaine Valley. This displays excellent columnar jointing and elongated vesicle trains (Fig. 5.8). Augite and olivine phenocrysts are to be seen in the compact core of the flow; the margins are scoriaceous.

6. Cross over the main road and ascend into Old Royat. There is a lovely old square here, good shops, several bars and the eleventh-century church of Eglise St-Léger, which is well worth a visit. Lunch can be taken here; there are seats in a small square for a picnic and several bars.

Figure 5.8 The fine old church of Eglise St-Leger in Old Royat, perched above the massive columnar-jointed flow from Petit Puy de Dôme.

7. Pass the east side of the church and follow the path down towards the wall, from where there is a good view across the valley and towards Clermont-Ferrand. Descend the stepped path to the Grotte de Laveuses. This is a lava tube within the basalt flow (it can be followed across the other side of the road too). Here the porphyritic lava is columnar jointed and shows excellent vesicle trains. The jointing gets better to the east of the grotto. It is also possible to see evidence of laminar flow within the flow, picked out by regions of rougher and smoother weathering on the outcrop.

8. Climb across the valley and up to the main road (D 68). Now walk up-hill for about 250 m, to the last house on the right, immediately past which is an indefinite path that rises steeply, straight up the hillside. Take it and encounter outcrops of granite with large K-feldspar phenocrysts. These weather out and can be collected readily. At the top of the steep climb, a path crosses the route; follow the sign to les Graviers.

9. On reaching the top of the plateau you enter verdant meadows – filled with wildflowers in spring. There are few rocks to be seen, but the views are good. In about 0.2 km bear left (ignoring the path that continues straight ahead) and cross the meadow diagonally, leftwards, towards the trees. Pick up another path that leads into Villars. The trackside drystone walls contain superb specimens of porphyritic lavas. Eventually you will cross the rim of the Maar de Villars once more, and enter le Cheix. Continue to la Baraque.

Excursion 3
Northern maars, Strombolian cones and flow fields

Introduction
The most northerly volcanic feature of Puy-de-Dôme is the Gour de Tazenet. South of this are several maars and some Strombolian cones, together with flowfields associated with the more northerly centres. This excursion investigates a diverse region in which quarries provide excellent sections through various cinder and scoriae cones; it takes a full day.

Starting point
Volvic, town square.

Itinerary
The town centre is a good place to start this tour, since most of the buildings are constructed of the dark Volvic stone – a compact well jointed trachyandesite that emerged from Puy de la Nugère about 10 000 years ago. This stone has been used for building since the thirteenth century. Volvic town is located on the edge of lava fields created by these eruptions. The church – at the back of the square – is well worth a visit; although restored much more recently, parts of it date back to the twelfth century, when it formed a part of the Priory of Mozat. Naturally, it is made of Volvic stone.

1. Leave the town centre and drive towards the northern end of town, whereupon a steep road forks off on the left-hand side and quickly climbs to Château de Tournoël. This great battered stone relict from the Middle Ages is set on a granite knoll on a fault splinter extending northeast from the western border fault of the Grande Limagne, and affords superb views over the plain of the Allier. The Limagne hills lie in the distance, and the faulted scarp that bounds the Limagne on the west, is clearly visible. So also are some of the cones and flows that spread out onto the floor of the Allier Valley. (The castle is one of the most interesting in Auvergne and can be visited at most times of the year; the entry fee is FF20. Tel. 73 80 58 86).

2. Retrace the route through Volvic, driving south on the D15 until it joins the D986. Turn right and continue along the edge of the Nugère flow until a sign indicates a right turn to Maison de la Pierre. Drive along the narrow road and park outside the entrance. (The same point can be reached by car or on foot only, by taking the narrow, twisting and steep road that leaves immediately south of the square. Beware: a coach cannot negotiate the bends on this short cut.) This tourist attraction was opened at the instigation of the Parc Regional du Volcans, and was installed in the first hundred metres or so of a lava tube that has been

enlarged by mining for the famous stone, a process that continued for 700 years. At the entrance is a good bookstall where geological maps can be purchased. Maison de la Pierre opens between 15 March and 15 November, between 10 h and 11.30 h and between 14.15 h and 15.45 h. It is closed on Tuesdays. Guided tours (all in French) take place every 45 minutes (fee FF20) and explain the geology and history of quarrying and building in the Volvic area.

3. Beyond the visitor centre at Maison de la Pierre is a quarry face in the Nugère flow. Small phenocrysts of augite can be seen, and the jointing is well developed.

4. Return to the main road and turn right towards le Cratère. On reaching the D941, turn left and then almost immediately right before the railway. Drive about 200 m and pull off into a large parking space on the left-hand side. Ahead lies the huge quarry face cut into the east flank of Puy de la Nugère. Until recently it was possible to get hands on the face, but now it has been fenced off. Nevertheless, it is well worth the trouble to inspect the bedded scoriae and ashes from the fence (Fig. 5.9). One can pick out many whitish blocks within some bands; these include charnockites from lower crustal levels. There are smaller numbers of peridotite nodules too. Small samples of both may be collected from the parking lot or the scree on the right-hand side of the face. These massive deposits belong to the pyroclastic mantle generated during an explosive trachyandesite phase at Nugère, which was accompanied by

Figure 5.9 Bedded scoriae with charnockite and other blocks, quarry at Gare-de-Volvic, Puy de la Nugère apron.

formation of a lava lake within the central crater. The edge of the latter lies some 400 m to the west, and is heavily vegetated.

5. Continue along the road to the north and inspect roadside outcrops of the trachyandesite flow. In about 450 m, the road turns sharply under a railway bridge. It is possible to park on the track that leads straight on (i.e. leaves the road on the left). Ahead, a marked footpath (GR441) will be seen; this can be followed gently upwards through the trees for about 500 m, whereupon one enters one of the old quarries exploiting the Nugère trachyandesite (there is some industrial archaeology here).

6. Return to the road and continue under the bridge. Soon, the D 90 joins the D 16, which leads through Moulet Marcenet. Immediately upon leaving the village, the cone of Puy de Paugnat rises prominently from the plateau, to the left. A huge quarrying operation has removed much of the cone's northern side. This cone, Puy Guettard and Puy Gonnard form a trio of very typical scoriae cinder cones built up by Strombolian-type eruptions. That is to say, the pattern of eruption is that of pulsatory explosions, some large, some small and others – more occasionally – major, that build up steep-sided conical structures with exterior slopes of 30° or so (Fig. 5.10a). Normally such a centre will have a summit crater or, if several separate eruptive phases have occurred, a cluster of craters, sometimes concentric and sometimes eccentric (e.g. Puy de Pariou and Puy de Côme). The cone-building periods may be punctuated or accompanied by the extrusion of lava flows.

In contrast and as exemplified by Cratère de Beaunit (this excursion) and Narse d'Espinasse (Excursion 5, pp. 54–58), on its way to the surface magma may encounter groundwater, which is converted into steam instantaneously, causing a violent explosion. The explosion punches a more-or-less circular crater in whatever basement is involved, throwing out blocks and debris around the crater as a low rim. In this way, maar craters are generated (Fig. 5.10b). The contrasting form and eruptive style is nicely revealed on the ensuing part of this excursion.

7. *Puy de Paugnat* Some years ago this quarry was accessible and provided a good opportunity to see a variety of large volcanic bombs, as well as collect exotic enclaves from within the scoriae sequence. Currently, access is denied. However, it may be worth checking, in case things have changed. It may be approached by following the D 90 just past the turn to Paugnat village, and taking a track up through the trees on the left.

Continue westwards along the D 90 until, in about 1 km, a road turns off on the right (signposted to Beaunit). Take this and continue north for about 1.5 km. An opening with a bar across the entrance will be seen on the right. Park here and enter the quarry.

(a)   (b)

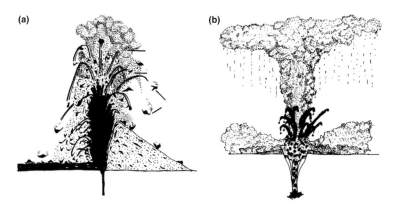

Figure 5.10 Formation of (a) Strombolian scoriae/cinder cone and (b) maar crater (© Pierre Lavina).

8. *Puy Guettard* (also known as Suc de Beaunit)   It is possible to collect mantle-derived peridotite nodules and some weathered granitoid enclaves from both the scree and walls of this quarry. (It is also a good picnic spot on a hot day). However, take care when climbing the talus slope as the face above is decidedly crumbly.

Return to the road and continue north, passing through the hamlet of Beaunit. This is located on the floor of a 1 km-diameter maar crater, Cratère de Beaunit, whose low rim can be seen ahead. If farmyard and other mess allows, it may be possible to pick out the vesicular porphyritic basalt flow that occupies the ground west of Beaunit hamlet. It is sometimes seen just before the road crosses the stream. Climbing up out of the village, approach a crossroads and an obvious small quarry on the left-hand side. Park here.

9. *Rim deposits of Cratère de Beaunit*   An interesting face exposes well stratified basaltic volcaniclastics, some appearing to have been laid down in water, probably on the interior of the crater. Several of the larger blocks form sags; others are draped by overlying strata. The relationships between blocks and scoriae, and the nature of some of the stratification (cross stratification in places) provide scope for discussion. To the south can be seen the younger Strombolian cone of Puy Gonnard that built up on the southern part of the crater's floor. The events that built the maar and cinder cone are depicted in Figure 5.11.

Continue up to the road junction with the D138 and turn right towards Charbonnières-les-Varennes. In about 250m it is possible to walk up a path into a field that affords the best view across the maar and the cone of Puy Gonnard. A further 250m along the same road there is an opening on the right-hand side, across which is a chain. Park here.

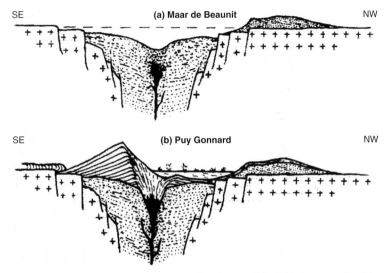

Figure 5.11 Evolution of Maar de Beaunit and Puy Gonnard. (a) Initial hydromagmatic activity generates maar crater with low rim. (b) Strombolian cone (Puy Gonnard) built on southern part of crater floor and ejected cinders drape maar rim. Finally, lake beds deposited in maar depression. (© de Goër de Hervé & Camus 1991.)

10. *Quarry in basaltic flank deposits of Maar de Beaunit*   This is an extensive quarry that exposes a variety of maar-related rim deposits, as well as ejected scoriaceous material associated with Puy Gonnard. Exotics within the maar sequence include large blocks of granite and some charnockite blocks (Fig. 5.12). On the west side of the face it is possible to discern dune and anti-dune stratification, probably related to an initial surge. The block-bearing part of the sequence is overlain by reddish cinders with few large blocks near the south wall.

Now take the the D16 northwards to Charbonnières-les-Varennes. Proceed through the town to the D227. Turn left and drive to Manzat. Now turn onto the D19 and drive northwards for about 2.5 km to Gour-de-Tazenat. There is a large parking area beside road.

11. Gour de Tazenat is the most northerly structure formed during the Puy-de-Dôme volcanic cycle. It measures 650 m in diameter and is about 60 m deep. Its age is something less than 10 000 years, although it is impossible to be precise. Take the short path from the parking lot to the lakeside where there is a small bar/restaurant (excellent crêpes) open from Easter to 1 October and where it is possible to swim or take a pedalo out on the water. The rocks near to the beach are rhyolitic tuffs which are injected by silicic veins. Now climb up leftwards to the rim, where there is an information board and viewpoint. One now has very fine view over the lake, the surrounding rim, and southwards towards

Figure 5.12 Bedded scoriae with angular granitic blocks. Quarry in rim facies, Maar de Beaunit.

Puy de Dôme. It is possible to walk right around the lake (takes about two hours), but there is little rock to be seen, except where the path descends to lake level.

12. Just before the information board is reached, on the opposite side of the road, is a small quarry that exposes well stratified breccia and tuff, with some granitic material. The violent explosion that formed this basaltic maar evidently brought up granitic rocks from the lower part of the crust, as is common with these highly explosive events. They were hurled out over the rim of the depression.

13. It is also possible to walk to a quarry in the southwest flank of Puy de Chalard, a typical Strombolian cone that is about 1 km distant. Simply return along the D 19 and take a left fork to a parking spot. A short walk leads into the quarry that exposes dark cinders.

Excursion 4
The area around Puy de Dôme

Introduction

With so many volcanic structures in one area, it is difficult to decide which are the more worthwhile to visit. Those selected are both accessible and easily linked into feasible itineraries with others that provide a contrast in style or aspect. It is entirely appropriate that the first excursion into the main part of the chain encompasses a visit to the summit of Puy de Dôme itself (1464 m), one of the finest viewpoints in France, if volcanoes are to your taste.

Starting point
Royat, or Orcines. Drive up the D 941a, to the junction with the D 68 near to Enval. To cover all of the localities and do them justice – including the walking ascent of Puy de Côme – it requires two days to make all the visits, without undue and undesirable haste. If the latter is omitted, everything can fit into one full day.

Historical background
The famous scientist, Blaise Pascal, spent much time in Clermont-Ferrand, not surprisingly since he was born there in 1623. In 1648, as a part of his studies of the atmosphere, he measured the difference of the height of a column of mercury with changing altitude. For this experiment he sited a measuring station at the summit of Puy de Dôme and another in Clermont-Ferrand. Local interest in the mountain led to the eventual completion of an observatory at the summit in 1876. It was during this operation that Gallo-Roman coins were found at the Temple de Mercure. The hideous modern television aerial and its accoutrements were built in 1957.

Since 1861 a railway has connected Clermont-Ferrand with Paris, and this was responsible for a not inconsiderable tourist industry based largely on well-to-do city dwellers. The Parisians were drawn to the mysterious mountain that rose abruptly from the plateau, but at that time it could only be reached on foot or by mule. However, in 1907 a tramway was installed; starting at Clermont, it passed through les Quatres-Routes, Durtol, le Grand-Tournat, la Baraque and le Font-de-l'Arbre, before attaining the summit area. The entire trip took two hours up and only one hour down. A contemporary postcard illustrates the tram at la Baraque and features as the frontispiece of this book.

With the development of the motor car, a toll road was built in 1928, with a maximum gradient of 1:7.5. Another interesting piece of history involves the prize put up by the brothers André and Edouard Michelin in 1908: they offered FF100 000 to the first person who could fly from Paris to the summit of Puy de Dôme. On 7 March 1911, the pilot Eugène Renaux and meteorologist Albert Senougue duly made the trip in 5 hours 10 minutes, in a biplane with a 60CV Renault engine, at an average speed of 130 km per hour. Today the Paris–Clermont flight takes just over one hour.

Geology of Puy de Dôme
The mountain has a complex history. The first stage in its growth was the extrusion of trachybasalt lava and the building of a low pyroclastic cone

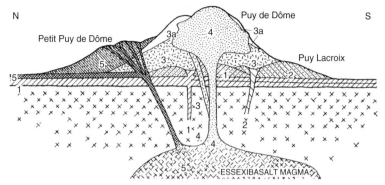

Figure 5.13 Cross section through Puy de Dôme, showing sequence of events that led to the present morphology. After various authors, including Wimmenaur in *The alkaline rocks*, H. Sorenson (ed.) (Chichester, England: John Wiley, 1974).

at a focus somewhere beneath the present structure (Fig. 5.13: 1). Thus, the obscure maar at Enval is overrun by leucobasalt flows (hawaiite) that can be seen spreading eastwards from beneath the main dome. The magma source that provided the first extrusion then differentiated by gravity segregation and crystal fractionation during a lengthy period of repose. This generated a stratified magma chamber in which cumulate essexibasaltic magma occupied the lower levels and was overlain by trachytic liquid.

The second phase of activity saw explosive eruptions and the growth of Puy Lacroix (Fig. 5.13: 2), whose Strombolian deposits are now largely covered by the younger products of Puy de Dôme. The third phase took the form of cone building, whereupon a steep pyroclastic cone of biotite–trachyte was constructed (Fig. 5.13: 3). The latter part of this stage ended with the explosive eruption of rhyolitic tuffs, which lie above the trachytic pyroclastics of the earlier part. During this highly explosive stage there was considerable loss of volatiles from the magma and a consequent increase in viscosity as time passed. The last fraction of magma associated with stage 3 was so viscous that it barely reached the surface from within the cone and it formed a massive plug, at some time between 10 800 and 12 000 years ago (Fig. 5.13: 4). Later, a catastrophic Plinian–Peléan event destroyed the eastern part of this dome, sending out rapidly moving nuées ardentes towards the east, probably about 9800 years ago. The scar left by this collapse is clearly visible today (Fig. 5.14). Subsequently a spine of viscous trachytic magma (locally termed **domite**) plugged the summit region, but this too was destroyed, sending out more glowing avalanches. Those that emanated from the late-stage Killian Crater have been dated at about 8300 years ago. The ignimbrites formed by these glowing avalanches can be inspected in a quarry near Col-de-Ceyssat.

Among the late-stage products from this centre are the trachyandesite

Figure 5.14 Puy de Dôme viewed from the west (near Beaune). Note the outlines of the old and new domes and the scar left by the first collapse.

of Petit Puy de Dôme (Fig. 5.13: 5), seen as a mixture of essexibasaltic magma and what little was left of the fractionated trachyte, and flows of similar composition that emerged from low down on the north side of the dome. The modern aspect of the mountain is the product of honing by normal erosion.

Itinerary
Join the D68 and at first ascend the southwest flank gently, reaching the toll booth (FF22). The road now climbs more steeply and tortuously around the northern side of the mountain (stopping is forbidden). Towards the top, outcrops of pale trachyte appear on the right-hand side of the road. Park in the large parking area. Here there is an excellent visitor centre that includes fine exhibits, book- and gift shops, toilets and a café/bar. It is worth spending at least an hour here (particularly welcome in poor weather).

1. *Puy de Dôme* The views from the summit region on a clear day are stupendous. The clockwise walk around the mountain is highly recommended. Note, before taking the circuit, that trachytic pumice outcrops on the northwest side of the summit region. On the west stretches the Ceyssat forest, with the valley of the Sioule beyond; then, to the north, comes Petit Puy de Dôme, the Grand and Petit Suchet, the distinctive double-crater Puy Pariou and the trachytic dome of Grand Sarcoui. This northerly group is geomorphologically diverse, for it includes both conical cinder and scoria cones, and viscous domes whose profiles are much more rounded.

2. On the eastern side (table d'orientation here) are fine views across the

plain of the Limagne towards the Livradois and Forez mountains. It is also a good vantage point for studying the Limagne fault scarp and the flows and cones that have spread out onto the floor of the Allier Valley. Eventually one reaches the remains of the Gallo-Roman Temple of Mercury. Interestingly, the walls are not built from local stone but from blocks transported from Puy Clierzou, some distance to the north; originally they were faced in exotic marble. This is another excellent viewpoint. More often than not, passing hang gliders provide added interest as they swoop past like huge birds of prey.

3. Descending now to the visitor centre, a fine southerly panorama opens up and can be viewed from a platform beside the buildings. Col de Ceyssat lies immediately below the steep southern face of the dome, and the almost military-precision line of cones extending down to the distinctive pair of Puy de Lassolas and Puy de la Vache – whose lava-flow dammed Lac d'Aydat – provides a classic case of volcanoes aligned along a fracture (Fig. 5.15). On a clear day, the Massif du Monts-Dore and the Plomb du Cantal are visible from here.

4. Walk just beyond the west side of the visitor centre to the top of a marked path. This descends down the side of the dome towards Col de Ceyssat. Take this path and walk tortuously down, past excellent expo-sures of jointed trachyte (domite) that form part of the first dome. Cross a road and then, in about 20 minutes, the col is attained and it is possible to inspect large blocks of augite-bearing lava that once were randomly distributed about the area, but which now have been bulldozed into one spot, at the top of the parking area. These were originally ejected from Killian Crater which is located across the road from the col. (There is a bar here that serves drinks and meals; its opening hours are odd.)

5. Drive eastwards down the D68 (or walk down the same road) for about 0.75 km, whereupon a gap in the trees on the left-hand side leads into a small quarry (there is a parking place opposite, under the beech trees). Enter the quarry to inspect the leucocratic breccias deposited from **nuées ardentes** that rushed down this slope when the dome collapsed, 8300 years ago. The deposit is monolithic, being composed of light-coloured sanidine trachyte, and the blocks are quite angular and matrix supported. This rock represents some of the most evolved magma that developed during the Puy de Dôme volcanic cycle (65.7% SiO_2).

Leaving the quarry, retrace the route back up to the col and continue driving along the D68. This enters deeply wooded countryside, then emerges into open fields just before Ceyssat village. From here there is an excellent view back to the Puy de Dôme cumulodome that rises serenely from the plateau surface. Pass near to Ceyssat and take the D52 to Champille, then take a right turn (D559) towards the northeast.

49

Figure 5.15 View down the southern part of the volcanic chain from near the visitor centre. Note outcrops of pale domite in the foreground.

This quiet road passes close to the west flank of Puy de Côme.

6. *Puy de Côme*   About 1.5 km prior to the junction with the D 941b maps show a double dog-leg in the road; about 250 m before this point, marked footpaths head into the woods on either side. Park here. Take path 959, which leaves the east side of the road and heads around the southern flank of the huge cone of Puy de Côme. When due south of the summit, take a left fork that heads steeply up the southwest slope, eventually emerging out of the trees onto the grassy rim of the outer crater. Puy de Côme is a fine example of a Strombolian volcano that experienced at least two major phases of cone-building eruption. The first-encountered crater is the older; this is some 300 m across. Inside it lies the smaller, younger crater (7700 years old). The eruption from this centre also involved effusion of extensive trachybasalt (hawaiite) flows that spread westwards as two huge tongues, one extending to Pont-aumur towards the northwest, the other to Mazaye in the west about 11 000 years ago. It is possible to walk right around the summit region, whence there are fine views towards these flows and also the complex cone of Puy Pariou to the east, the spine of Puy Chopine towards the north, and the dome of Grand Sarcoui to the northeast (Fig. 5.16). Take the steep descent down the southeast slope and return to the parking spot beside the D 559. The walk takes in all about 2.5 hours.

Continue to the junction with the D 941b and turn left. In about 500 m a sign will be seen for Volcan Ouvert au Ciel. Pull off into the large car-park on the right-hand side. This tourist attraction was developed in the extensively quarried cone of Puy de Lemptégy and it is a very fine excursion. The entrance fee is FF12 per person. For groups the place is open all year; at other times the opening hours vary and it is advisable to phone ahead of a visit to check details. On entering the reception building one is ushered into a small theatre where a video presentation is made (about 20 minutes, all in French), but it is as well to go straight through to the quarry and start the unaccompanied tour, which involves following a sequence of very informative boards describing the geology at many points along an anti-clockwise circuit of the quarry. The descriptions of the geology are all in French, but each information board is illustrated with diagrams and it is entirely possible to pick up the story. A cross section through the cone is shown in Figure 5.17.

7. *Puy de Lemptégy*   Puy de Lemptégy is a Strombolian cone that has been deeply dissected, largely by quarrying. During the circuit one sees cinders and scoriae associated with cone building, dyke intrusion, lava flows and bombs of various kinds, together with lavas from the core of the vent (Fig. 5.18). The sequence is well stratified, and superimposed

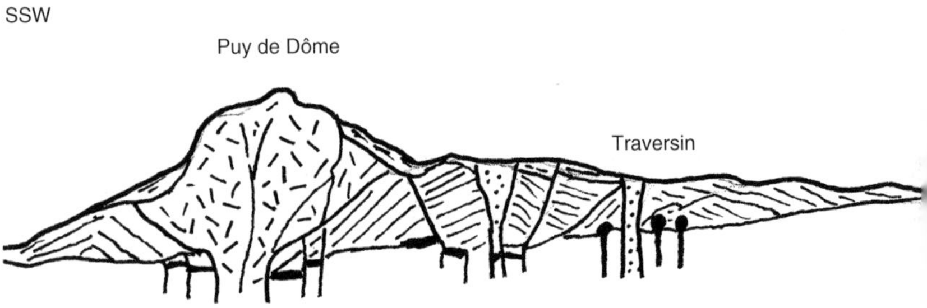

SSW

Puy de Dôme

Traversin

Figure 5.16 Above and opposite: Puy de Dôme to Grand Sarcoui.

on the Lemptégy rocks are deposits associated with the explosive eruption of nearby Puy Chopine, including possible laharic breccias and trachytic ashflows. Some of the interpretations are, to my mind, arguable, but this only adds to the interest of the place. It is a very worthwhile trip and one needs a good two hours to do it justice. Leaving Puy de Lemptégy, drive east along the D 941b towards Col des Goules. At the junction with the D 559, on the left-hand side is a geological information board that is worth investigating. It describes the evolution of this part of the volcanic chain and has a nice cross section of the geology. A gap in the trees affords views of the spine of Puy Chopine and the adjacent Strombolian cone, Puy des Gouttes. These two volcanoes are now entirely cloaked with forest and there is little point in making a serious hands-on geological tour, rock outcrops simply cannot be found in sufficient abundance to justify the time.

8. Turn left at the junction and drive along the D 559 towards the north-east. In about 2 km the road forks. Park under the trees on the verge. A track leads off the junction, in a southerly direction; follow this gently

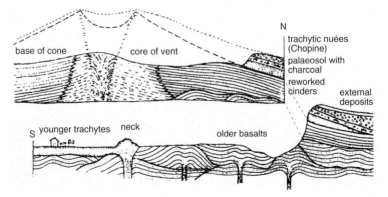

Figure 5.17 Cross section through Puy de Lemptégy (© de Goër de Hervé & Camus 1991).

NE

Puy Pariou   Grand Sarcoui

Col des Goules

uphill. A sharp right-hand turn leads into a quarry cut into the lower
southwest flank of Petit Sarcoui, where there are good exposures of
well jointed grey phonolite; there are also some volcanic bombs.

9. Continue walking east along a path that crosses the col between Petit
and Grand Sarcoui. The convex slope of Grand Sarcoui dome rises
steeply on the right-hand side. Eventually the path drops down steeply
towards the D 559, but just above the road a track leads into a very over-
grown quarry. A few years ago it was possible to stand back and see
leucocratic ignimbrite (similar to that on Puy de Dôme) and observe
dune stratification associated with ground-hugging surges. Now,
however, the birch scrub has quite taken over and you have to fight to
get up to the quarry face at all. If you have the resolve, there are quite
good outcrops of pale domite (trachyte), angular blocks being set in
something resembling dry cement powder.

Descend to the road through the trees, and pick up transport if it has

Figure 5.18 Party following the geological trail across the floor of Puy de Lemptégy quarry.

been sent round to meet the party. Alternatively, you can walk back to pick up transport left at the beginning of this walk; it takes in all about 15 minutes. Now drive east along the D559 to Ternant village. Ignore the left fork and carry straight ahead between the houses; go ahead again at the next junction (D559), but, just past the village, turn off this road to the right along the D90 (signposted to Orcines).

10. In about 300 m this road leads to a signposted viewpoint and large parking area on the left-hand side. It is a good spot from which to take in the fine panorama to the east, over the Allier Valley. Opposite the parking area is a small quarry that is also worth a brief visit. At the lower left-hand corner of the quarry opening are bedded red cinders. Moving into the opening, these give way to solid basalt that in places has developed weak columnar jointing. This is one of the few readily accessible exposures of the basalts erupted prior to the main Chaîne des Puys activity. The road now continues to Orcines.

Excursion 5
The southern area of the Chaîne des Puys

Introduction
This excursion provides the visitor with an opportunity to explore the more southerly part of the chain, with superb views from the summits of either Puy de la Vache or Puy Lassolas (or both). The nearby Maison du Parc des Volcans is well worth a visit and has excellent displays explaining the geology, botany and culture of the region; it also has a good bookstall. Passing through the lava-dammed Lac d'Aydat (where there is an opportunity to swim in the lake and there are also bars and restaurants), the route turns west past Puy de la Rodde and on to the small hamlet of Espinasse and its interesting maar (Narse d'Espinasse) and the associated Strombolian cone, Puy de l'Enfer. The route then meanders through lovely countryside, eventually reaching Chambon-sur-Lac, a convenient and extremely pleasant base for exploring the Massif du Monts-Dore.

Starting point
Royat–Orcines.

Itinerary
Take the D941a southwestwards towards Col de Moreno, then just under 2 km after Enval, fork left along the D767a and continue to Laschamp. Here turn along the D52 towards Beaune. There are fine views towards Puy de Dôme along this road, especially at the point named Croix Parla on

the map. Pass through Beaune and, at the junction with the D5, turn right towards Randanne. The road enters wooded countryside and in about 3 km reaches a large parking area with picnic tables, located at the base of Puy de la Vache. Park here.

Puy de la Vache and Puy de Lassolas
These two breached Strombolian cones are believed to have erupted at much the same time, along with Puy de Barme, constructing cones 1 km in diameter with 30° slopes that rise to around 1100 m. The fresh appearance and the clean black and red cinders that characterize these centres suggest that they are relatively recent, probably having erupted between 8500 and 7500 years ago. At the same time as the cone-building activity, both centres extruded flows of basalt that extended eventually as far as St-Amant-Tallende, a distance of about 16 km. In so doing they dammed several valleys, forming Lacs d'Aydat and Cassière.

1. Take the track that crosses the road and leads westwards from the car-park. In about five minutes a quarry is reached, on the right-hand side. Enter the quarry and inspect lava blocks, black and red cinders, and various types of bombs. The rock is a dark grey **augite-phyric** basalt.
2. Cross the quarry floor and follow a marked footpath steeply up through the trees. This follows the southwest ridge of Puy de la Vache and, although steep, is a good path with (ostensibly helpful) steps. It takes about 25 minutes to reach the summit area. From here a superb panorama greets the observer (Fig. 5.19). The breached craters of both

Figure 5.19 Ascending the cone of Puy de la Vache. The southern part of the volcanic chain is superbly seen. In the distance is the snow-capped Massif du Monts-Dore.

cones are now clearly seen, and on a clear day the water of Lac d'Aydat can be seen sparkling towards the southeast, the wooded basalt flows associated with the cone stretching out eastwards towards St-Saturnin. There are equally fine views to the north.

3. The rocks at the summit make a fine photograph and provide useful foreground for panoramic shots. To now return down the southwest ridge of Puy de la Vache, follow the path around the rim, then take a steep and somewhat cindery path leftwards and steeply down the side of the cone, eventually reaching the track at the base. Take care on this descent because the ground is tricky if wet. Now return to the car-park. If energy abounds and it is thought a good idea to take in Puy de Lassolas too (well worth the extra effort), walk westwards from the summit rocks until a path is spotted that leads northwards down to a deep col and then steeply up again to the summit of Puy de Lassolas. To reach the summit it takes another 25 minutes. The views from here are even better than on top of Puy de la Vache. Continue around the rim and follow a well trodden path down to the track to the car-park.

4. Continue along the D5 towards Randanne. Blocky lava can be seen along the roadside at several places. Shortly, Maison du Parc des Volcans will be seen on the right-hand side. Turn out and park up (toilets here). An extremely interesting hour or so can be spent here; the exhibits are well set out and imaginatively documented. There is also a bookstall that sells a variety of books, maps and postcards.

5. Leaving the visitor centre, continue to the crossroads at Randanne. Turn left along the D89. In a short while a large parking bay on the left-hand side leads to a fine view back towards Puy de la Vache and Lassolas and, indeed, to Puy de Dôme and beyond (Fig. 5.20). There is

Figure 5.20 View from the information board on Excursion 5. Puy de Dôme is prominent in the centre.

a very good information board here that describes the geological story of these centres and those adjacent.

6. At Col de la Ventouse, turn right (south) towards Lac d'Aydat along the D 213. Now take the D 5 (to the right) and enter the small village of Fontclairant. There is a parking spot on the left. Walk on about 50 m and cross over the road to a small cutting that exposes basement granite overlain by 2 m of cinders and pale ash. Carry on to the road junction and turn left towards Lac d'Aydat.

7. Leave the village via the D 5 (signposted Zanières) and note red cinders and black basalt on the roadside. Almost immediately turn right along the D 5E towards la Garandie. Pass through the village and at the T-junction turn left towards the hamlet of Espinasse. At the first houses note a track on the left-hand side (signposted to Narse); park here.

8. *Narse d'Espinasse and Puy de l'Enfer* These two landforms show activity of different types, simultaneously, along the same fissure. An initial stage of hydromagmatic activity to the south generated an explosion crater or maar through the underlying basalts of the Monts-Dore volcano. More or less simultaneously, Strombolian activity began immediately to the north, building the early cone of Puy de l'Enfer. Activity seems to have continued along these lines until a final massive explosion of the maar blew away the entire southern part of the Strombolian cone, which by this time had grown quite large (Fig. 5.21).

Walk along the track, gently descending to the floor of the maar

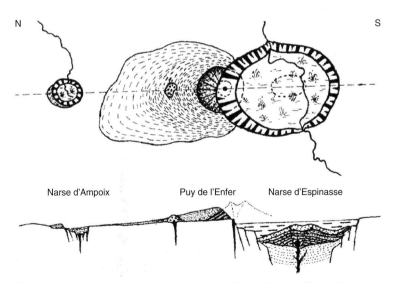

Figure 5.21 Sequence of events leading to formation of Puy de l'Enfer and Narse d'Espinasse (© de Goër de Hervé & Camus 1991).

depression. If the lighting is low enough, the rim of the crater can be traced almost in its entirety. Climbing out of the crater, ascend towards the cone of Puy d'Enfer. In a short while a track leads off to the right (chain across to prevent cars entering quarry); take this track and enter the quarry. Within the stratified quarry walls, products of several eruptions can be discerned. Bedded cinders belonging to the basal part of the original Strombolian cone, different kinds of bombs associated with the maar, and reddened cinders associated with the later stages of cone building, can all be seen. There are discontinuities between the different deposits. It is also a fine viewpoint across the maar depression and towards the Massif du Monts-Dore.

9. Return to the vehicle and drive on through Espinasse and into Saulzet-le-Froid. Now take the D74 to Zanières, bearing right towards the forested cone of Puy de Montenard. Fork right onto the D5 and drive to Chambon-sur-Lac. There is a fine view as one descends past Château Murol towards the valley floor. Halfway down the descent to Lac Chambon a large parking area on the right-hand side leads to a geological information board describing the inverted relief at Château Murol and in the plateau beyond.

Excursion 6
Col des Goules–Puy Pariou–Puy de Dôme–Col de Ceyssat walking tour

Introduction
This is a moderately strenuous but highly exhilarating excursion that traverses two of the main volcanic edifices of Puy-de-Dôme and provides the most wonderful views over the surrounding region. At a moderate pace the excursion should take about 3½ hours to the summit of Puy de Dôme and a further 30 minutes down to Col de Ceyssat. Assuming one refreshment stop, the entire trip would take five hours.

Starting point
At the information board near Col des Goules, on the D941b.

Itinerary
About 400 m to the west of the information board at the side of the main road, and on the opposite side of the road, a marked path leads through the trees across the edge of Plateau de Fraisse. This ridge is the rim of the maar crater that existed here before the more recent cone of New Pariou was built. After about 450 m, the path forks; take the left-hand path and climb steeply up the sides of New Pariou.

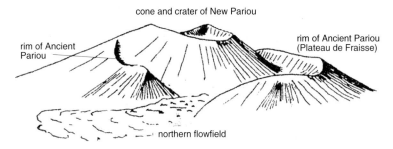

cone and crater of New Pariou

rim of Ancient Pariou

rim of Ancient Pariou (Plateau de Fraisse)

northern flowfield

Figure 5.22 Multi-stage eruptive history of Puy Pariou. An inner and outer crater complex were formed and a prominent flowfield, which left the breached rim of the older crater along its northern side.

1. As you approach the top of the cone from the northeast (i.e. after emerging from the woods), the crater complex becomes readily discernible (Fig. 5.22). From the grassy crater rim, you look down across the Chaîne des Puys. This volcano had a multi-stage eruptive history:

   (i) Ancient Pariou was a basaltic scoria cone formed more than 9000 years ago by Strombolian-style activity. Only a small part of this can still be seen, low down on the west side of the edifice.

   (ii) Subsequently the focus of activity moved slightly towards the northeast, and a major hydromagmatic event generated a large maar. This in turn was filled by a lava lake about 8000 years ago.

   (iii) Subsequently, a new cinder cone was built up (New Pariou), but this time the magma was a trachyandesite.

   It appears that, early in the growth of this edifice, the cone was breached by lava pressure and a substantial flow leaked out of the central depression, giving rise to the extensive flow that spreads out first towards the north, but then turns towards the southeast. This flow has been dated at 8200 years ago and is the one encountered on Excursion 2 (pp. 36–40). The new cone continued to grow after this event, and the current summit crater measures 200 m across and is 90 m deep.

4. It is possible to circumnavigate the summit crater, enjoying magnificent views all around, before descending obliquely down the western flank towards the col between Pariou and le Petit Suchet. Continue walking due south, crossing the vent of le Traversin and then, joining path GR4/GR441, ascend up and over Nid de la Poule, close to the summit of Petit Puy de Dôme. Both le Traversin and Nid de la Poule are the remains of basaltic maars, the hydromagmatic eruptions generating a "nest" of depressions in the latter case, well seen from the path ahead, which leads to the top of Puy de Dôme. These both erupted well before the emplacement of the adjacent cumulodome.

5. Now the steep climb to the summit begins, taking a more-or-less direct

route up the northern slope. Eventually the path intersects the road, whereupon follow this clockwise around the dome, until the top is reached. On the way up, good exposures of evolved trachyte lava (domite) are to be seen on the right-hand side of the road. (Do watch for traffic; this is a dangerous place.)

6. *Summit of Puy de Dôme*. See Excursion 4 (pp. 45–54) for details of the summit region and the descent to Col de Ceyssat. This excursion can, of course, be terminated at the summit car-park.

Excursion 7
Puy des Goules–Grand Sarcoui–Puy Chopine–Puy des Goutes–Beauregard

Introduction
This easy walking tour takes a full day and nicely illustrates the different kinds of volcanic landform encountered in the Chaîne des Puys. There may not be a huge amount of good exposure, but the geomorphology and scenery are extremely good. It is advisable to arrange for transport to drive to Beauregard (just south of the D941 Pontgibaud–Volvic main road) to pick up at the end of the excursion.

Starting point
Col des Goules, beside the D941b.

Itinerary
Just east of the Col des Goules, there is a break in the trees on the right-hand side of the road. Take the footpath that skirts the edge of the forest until, in about 400 m, it forks. Keep left, left again at the next fork and gradually climb up through the trees towards the col between Grand Sarcoui and Puy des Goules.

1. *Grand Sarcoui* In about 500 m it is possible to take a path right that leads to the caves on the southeast flank of the cumulodome of Grand Sarcoui and thence to the summit region. The domite surrounding the caves is a friable leucocratic material out of which weather many biotite flakes. This rock has been widely used for glass making. Note the convex form of the dome and the flat top typical of such structures. This contrasts sharply with the adjacent Strombolian cone that the route next ascends.

2. *Puy des Goules* Descend by the same route to the col, then turn right, towards the west. In about 200 m a track climbs steeply up through the trees on the left-hand side. Follow this to the summit crater of Puy des Goules. This Strombolian cone has a fine crater with a depth of 30 m.

Excellent views towards Puy Pariou, Grand Sarcoui and Puy Chopine are to be enjoyed from this point. It is possible to walk around the entire rim before descending *by the same path* to the col (any other route leads into thicket and is not to be recommended).

3. *Puy Chopine and Puy des Gouttes* Rejoin the path across the col, turning left (west) then, in 500 m, join a track and turn left. Follow this for 350 m, before turning right along footpath GR4/GR441. Follow this through fairly dull plantation for a further kilometre before taking a left fork (GR441 to Puy Chopine).

ALTERNATIVE 3A: At this point one can go straight ahead and reach the summit of the spine of Puy Chopine via a forestry track. This is not difficult, although steep, but the entire route is encased in old forest, so you see little; however, there are some exposures of the pale trachyte beside the track. If this route is taken, on descending the spine to the main footpath, turn left and continue around the northwest flank of the structure until rejoining GR441 and the track to Beauregard.

ALTERNATIVE 3B: Instead of going straight ahead at the junction, I would recommend bearing left and climbing up through thickly wooded terrain to the col between Puy des Gouttes and Puy Chopine (Fig. 5.23). This has the advantage of being less strenuous and, furthermore, this route affords nice views of the sharp spine of Puy Chopine, which sits within the breached crater of Puy des Gouttes, the summit of which lies to the left of the path. (In my view, like Puy Chopine, the ascent of either puy is not worth the considerable effort.) The earliest phase of activity in this area was the construction of the basalt scoria cone of Puy des Gouttes. Subsequently, a massive explosion blew out a crater at least 600 m in diameter and generated nuées ardentes that flowed down the northwest and eastern flanks of the volcano. The final phase saw the emplacement of the viscous trachyte spine of Puy Chopine, an event that occurred around 8200 years ago.

4. On reaching the T-junction, turn right along a track for about 350 m (at which point route 3a joins), here turning left along another track (signed GR4). Follow this through the woods; it twists and turns before reaching the lovely rural hamlet of Beauregard, where transport should have been left. It is then possible to drive to the main D 941 (Pontigibaud to Volvic road), reaching this at le Vauriat. There is a good les Routiers restaurant here that serves hearty meals and refreshing drinks, all day.

Figure 5.23 The spine of Puy Chopine (covered in trees) sits within the breached crater of Puy des Gouttes. Beyond and to the right is the massive cone of Puy de Côme, and Puy de Dôme lies in the centre of the distant skyline.

Excursion 8
Royat–Puy de Gravenoire–Gergovie–Montrognon–Veyre-Monton

Introduction

This is a mixed bag of an itinerary. It begins with a look at a large cinder cone, passes through granite country, investigates older volcanic necks and basalt flows, visits Oligocene sedimentary rocks and provides those who follow it with two of the finest views across the Limagne Valley from anywhere in Auvergne. The volcanicity in this region spanned the period between the Stampian (Upper Oligocene) and the Pliocene, the rock types ranging from ankaramites, through basanites, basalts, trachyandesites to trachytes. They also include interesting mixed volcani-sedimentary rocks called pépérites in which somewhat rounded vitreous lava fragments are enclosed in a calcareous matrix. They are believed to have originated in the forceful injection of either trachytic or basaltic magma into marly sediments, the rapid cooling and degassing causing explosive disruption of the lava, producing the observed clasts that became intrusively injected into the sedimentary strata.

Starting point
Royat.

Itinerary
Leave Royat on the D941a to Ceyrat. In about 1 km turn right along the D767 and ascend towards the prominent cone of Puy de Gravenoire. Arrive at a T-junction.

1. *Puy de Gravenoire* This used to be one of the most frequently visited of the northern puys, largely because of its proximity to Clermont-Ferrand. It is a typical Strombolian basalt cinder and scoriae cone, and it rises prominently from the granite plateau above the towns of Royat and Ceyrat. It has been extensively quarried for its cinders (known locally as **pouzzolane**) and its scarred shape is very distinctive. There are several large quarries that may or may not be accessible. Two of these lie to the right of the road junction, and several more lie to the left. It is worth trying all of them, as the red and black cinders are finely bedded, and excellent bombs, of both breadcrust and spindle form, may be seen. The compact basalt, when not entirely vesiculated, contains small phenocrysts of mainly augite, with lesser amounts of olivine and feldspar. If a left turn is taken at the T-junction, drive for about 1.3 km and bear left again; the road bends back on itself on its way down to Boisséjour and Ceyrat. At a huge hairpin bend, the largest quarry of all springs into view. This is approachable via the D941c,

which runs from Boisséjour to Royat, but is now largely given over to housing development. There is a very impressive exposure here of a 200 m-high fault-line scarp bordering the Grande Limagne.

2. *Montrognon* Drive through Boisséjour and join the main N 89 road. Turn right (south) towards Ceyrat. In the town take a left turn signposted to Montrognon. Drive as high as possible towards the feudal castle and ruined church that caps this old volcanic boss, and park up. Walk up the hill towards the castle and note outcrops of the pépérite. The boss is largely basaltic but overlies the calcareous marls of Stampian age to be seen later on this excursion. This is one of the older necks that pre-dates the puys of the main chain. At the castle (which is well worth the visit) there are fine views over the plains below. There are also some splendid rock specimens to be seen in the surrounding walls.

3. *Plateau de Gergovie* Descend to Ceyrat and take the N 89 road towards Saulzat-le-Chaud. Turn left (road signposted Gergovie) and drive to the summit of the Plateau de Gergovie. There is a large parking area and restaurant here. Gergovie, legend has it, is the spot where the local hero, Vercingetorix – whose ambition was to rid Gaul of the hated Romans – had a famous victory over the troops of Caesar in 52 BC. Whether or not this is true, thousands of tourists visit this place every year in homage to the local man. It is the remnant of a lava plateau and it rises some 400 m above the Allier River. Beneath it lie Oligocene marls and limestones.

The plateau is capped by two basalt flows pre-dating the Chaîne des Puys, which were emplaced on a surface of Stampian (Oligocene) sedimentary rocks (Fig. 5.24).

The upper flow has an age of 16 million years, the lower, 19 million years. Drilling has revealed these sediments to be at least 1750 m thick near Riom. Regrettably, vegetation has covered most of the old exposures; however, some basalt can be seen in the vicinity. The main point of the visit is the exceedingly fine and wide view the plateau edge affords of the Limagne Valley and the Forez Mountains beyond.

4. *La Roche Blanche* Return by the same route and turn left. After 2 km or so is Varennes, where another minor road leaves the N 89 road on the left-hand side. Take this and drive through Chanonat and reach the old hilltop town of la Roche Blanche. The limestone is extensively evident here. The town is something of a maze of narrow streets; however, shortly a signed road to the village of Gergovie should be spotted on the left-hand side. Take this and park up in the little village, near the bar.

In Gergovie village, go on foot from Place St-Jean along Rue de la Cure, past le Plat and the last houses. The road is narrow and winding, and it is easy to get lost: do persevere.

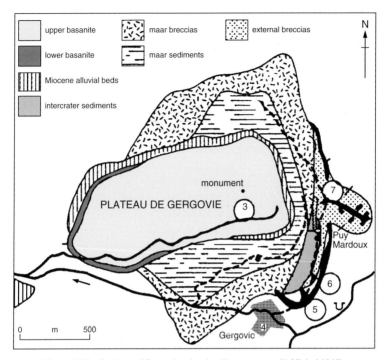

Figure 5.24 Geology of Gergovie, showing itinerary stops (© Michel 1948).

5. *Road along at southeast corner of plateau* Follow the gravel track running under the southern face of Gergovie plateau. Take this, noting the occasional outcrop of white marl, and in a short distance see the marl cut by a thin cone-sheet of basalt.

6. *Small quarry at two levels* Soon a small quarry will be found on the left-hand side; this is at two levels. White marl of Stampian age is to be found here, together with **apophyses** of olivine basanite and some pillow-like lava.

7. Continue along the road for a further 250 m. One soon encounters a large mass of pépérite in which calcareous marl acts as a matrix for dispersed clasts of vitreous lava. Further on again is Puy Mardoux, made from the same stuff but here intruded by basalt dykes and sheets.

8. Leave the little village and carry on eastwards along the D 52e; this road passes over the motorway and intercepts the wide D 970 just over the other side. Turn right along this – it is a fast straight highway – and continue along it for about 3 km, whereupon a road on the right leads, extremely tortuously, to the hilltop town of Veyre-Monton. The road climbs and climbs until you are among high buildings that seem almost to touch each other. You will pass the church and continue onwards

until it seems you can get no farther. In due course a turn up right, very steep, is signed to a viewpoint. It is worth the seemingly hazardous drive for two reasons:

- There is the most extraordinary and bizarre monster of a marble statue here.
- There is one of the finest views across the region, a real full-circle experience. It is a good way of spending a few final moments before driving to the airport at Aulnat, a mere 20 minutes to the north.

Chapter 6

The Massif du Sancy

Development of the Dore–Sancy volcano

The mountains surrounding Puy de Sancy are undoubtedly among the most impressive in Auvergne. Viewed from the spa town of le Mont-Dore, Sancy presents a formidable heavily glaciated face, crowned by a rugged summit from which radiate steep-sided ridges. Puy de Sancy (1885 m) and Puy Ferrand (1854 m) are the two more prominent tops and are joined by a high ridge. On the north side, a deep amphitheatre has been etched out by the combined action of water and ice, giving rise to the narrow U-shape valley through which the Dordogne River now flows and on whose floor sits le Mont-Dore (Fig. 6.1). To the east, where the Chaudefour Valley and Lac Chambon are prominent landforms, similar erosive forces have been at work, and on the west is the Plateau de l'Artense, to the south the lovely lakes of Chauvet and Pavin, and to the north the plateau of Ordanche and Lac de Guéry.

This impressive massif is the dissected remnant of a large stratovolcano (in truth, there were three that coalesced) measuring 33 km north–south and 16 km west–east and extending over an area approaching 650 km². Its eruptive history is complex, involving the emission of lavas of a wide range of composition, explosive ejection of pyroclastic airfall and ashflow deposits and intrusion of a plethora of dykes, viscous plugs and spines. The products of eruption occupy a volume variously estimated at 200–250 km³, and its growth spanned the period from the uppermost Miocene (7 million years ago) to the Pleistocene, with localized episodes yielding ages as recent as 12 000 years. In its heyday this volcano would have surpassed Vesuvius in both its scale and grandeur.

Activity in the region began about 7 million years ago with eruption of basaltic lavas from fissures that fractured the brittle granitic basement. Remnants of these early flows are to be found east of the Sancy massif, near Murols, around Roche Romaine, Olloix and Grandeyrols. The majority of these early flows filled pre-existing valleys in the basement; however, subsequent erosion has left the old lavas, which became buried by a great

Figure 6.1 The snow-covered summit of the Sancy massif, from near Lac Chambon.

thickness of pyroclastic rocks, as prominent plateau-like areas of inverted relief, their having stood firm while the friable volcaniclastic deposits were inexorably stripped away. There seems little doubt that these initial outpourings represent a continuation of the Miocene activity that gave rise to flows in the valley of Limagne, to the east.

About 3 million years ago, another pocket of magma rose into the granitic crust, generating fractures while doming it, and then explosively disgorging large volumes of rhyolitic pumice and ash. The emptying of this second magma chamber eventually led to a major collapse, giving rise to the volcanotectonic depression of Haute-Dordogne, a large fault-defined caldera structure measuring 12 km by 10 km. For about a million years, a sequence of predominantly Peléan and Vulcanian explosive erup-tions gradually filled it with pyroclastic rocks. Where ephemeral lakes formed, these filled with a variety of sedimentary rocks that became inter-bedded with the volcanic deposits. This activity gradually built up into the stratovolcano known as that of Monts-Dore. Today, at the periphery of the old volcano the granite basement lies roughly 300 m higher than at the foot of the ancient caldera's fault-defined wall.

One of the commonest rock types produced during this explosive stage was termed by French geologists **cinérite**, a deposit of **indurated** cinders. Similar airfall deposits were not simply content to infill the depression but eventually spread beyond its boundaries. At la Bourboule, it is possible to identify a series of Lower Cinérites, which comprises a sequence of scori-aceous beds rich in detrital clasts, often light in colour and of relatively fine grain size. These are overlain by a group of Upper Cinérites, which are much more blocky and include some very massive breccias. It was the latter series that contributed most to the construction of the Monts-Dore edifice. Included within the pyroclastic succession are Vulcanian airfall breccias, Peléan nuées ardentes, lahars and fluvial conglomerates, sands and muds, some of lacustrine origin. The lower series covers an area of 135 km^2; the upper series extends for 350 km^2.

Whereas cone building was largely achieved by explosive eruptions and pyroclastic flow episodes, lavas also were poured out within the depression and became interstratified with the pyroclastics; they also solidified in the form of domes and flow domes. For instance, the vitric rhyolite of Lusclade (Excursion 4a, pp. 84–87) was emplaced early in the history of the volcano, mantled by an explosion breccia containing massive rhyolitic blocks, which, in turn, was completely covered by the Upper Cinérites.

Between 2.0 and 1.5 million years ago, activity became concentrated largely to the north of the caldera, with extensive outpouring of basaltic rocks in the region of la Banne d'Ordanche (in reality, a separate volcano),

69

and the emplacement of spines of viscous phonolite such as those of Roches Tuilière and Sanadoire. During this phase, explosive, hydromagmatic and effusive eruptions occurred.

The true Massif du Sancy did not appear until about 850 000 years ago. This younger stratovolcano grew up on the southern part of the caldera floor and continued to develop for the next 650 000 years. Its construction began with the rise of a smallish magma chamber whose subsequent evacuation threw out a large sheet of pumice and related explosive products. The stratified pyroclastic blanket was followed by extrusion of thick pasty flows to the east of Puy de Jumelle, to the north of Capucin, around Montagne de Bozat and at Puy Gros. Between 500 000 and 250 000 years ago there was a new emission of pumice and cinders in the region of Chaudefour, while domes were extruded at la Perdrix and Mont Redon. Finally, about 230 000 years ago, a violent explosion occurred and, among other events, the large trachyandesitic dome that now forms Puy de Sancy was extruded. The construction of the stratovolcano had ended and it remained for glaciation to modify its shape. This major climactic phase saw intrusion of trachyte, phonolite and rhyolite and extrusion of identical lavas; trachyandesite and basalt were also erupted. The rock types sancyite and doréite (saturated intermediate lavas) are specific to this volcano. There were also large volumes of breccias, scoriae, ignimbrites and many lahars. This sequence of events is beautifully depicted on the information table located near the summit of Puy de Sancy (Fig. 6.2).

Between 20 000 and 10 000 years ago, the old mountain was etched out

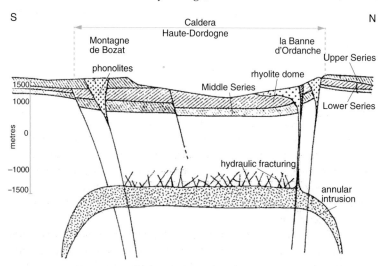

Figure 6.2 The evolution of the Monts-Dore volcano, as depicted on the Parc des Volcans information board near the summit of Puy de Sancy.

by erosion, largely via the work of glaciers that accumulated in the valleys of Haute-Dordogne, Chaudefour, Fontaine Salée and Chausse. The principal activity was associated with the Würm glaciation (12 000–10 000 years ago), when U-shape valleys were excavated and, where the 100 m-thick ice encountered obstacles such as flow domes and thick flows, it flowed around them, leaving an erosional path manifested in a collar-like feature. It also left many of the older low-lying flows as areas of inverted relief. After the ice retreated (and in some cases while it did so), there was a further spate of minor eruptions, particularly at Puy de Tarteret (12 000 years ago) which barred the valley of Chaudefour and generated the beautiful Lac Chambon. Large cones were also emplaced at Puy de Montenard and Puy de Montchal. Glacial erratics and morainic deposits were another legacy of this stage. A simplified geological map and a cross section of the region are shown in Figure 6.3.

One of the more difficult things to keep in mind while travelling the region is the size and position of the old Haute-Dordogne (Monts-Dore) caldera depression. It is useful to be able to do this, since it enables you to place the observed rocks in the context of the palaeogeography. Figure 6.4 shows the boundary of the Haute-Dordogne and Sancy calderas with respect to the main cross-country routes.

Rock associations of the Monts-Dore

Although at first sight there seems to be a bewildering array of rock types associated with the volcano, in broad terms the rocks can be divided into two groups:

- a **saturated** series, comprising the rock types: ankaramite–olivine-poor basalt–trachyandesite (doréite)–quartz-latite (sancyite)–quartz-bearing trachyte–rhyolite
- an **undersaturated** series, comprising: olivine-rich basalt–leucocratic labradorite–leucocratic tephrite (ordanchite)–phonolite.

Analyses 1 and 2 in Appendix 4 represent the two end members. Differentiation of the olivine-poor variety led to development of rhyolites, whereas differentiation of the olivine-rich type produced phonolites. In both series, the plagioclase-rich rocks form lava flows, and the phonolites and silicic types build plugs. Pyroclastic rocks such as ignimbrites, pumice flows and rhyolitic tuffs are common in the saturated series, but not in the undersaturated one.

At this point, you might reasonably ask why two different rock series developed in the first place and if this had any significance for the history of the volcanoes. As a starting point, let us assume that the parent magma was an olivine-rich basalt. If this were allowed to crystallize in the normal

71

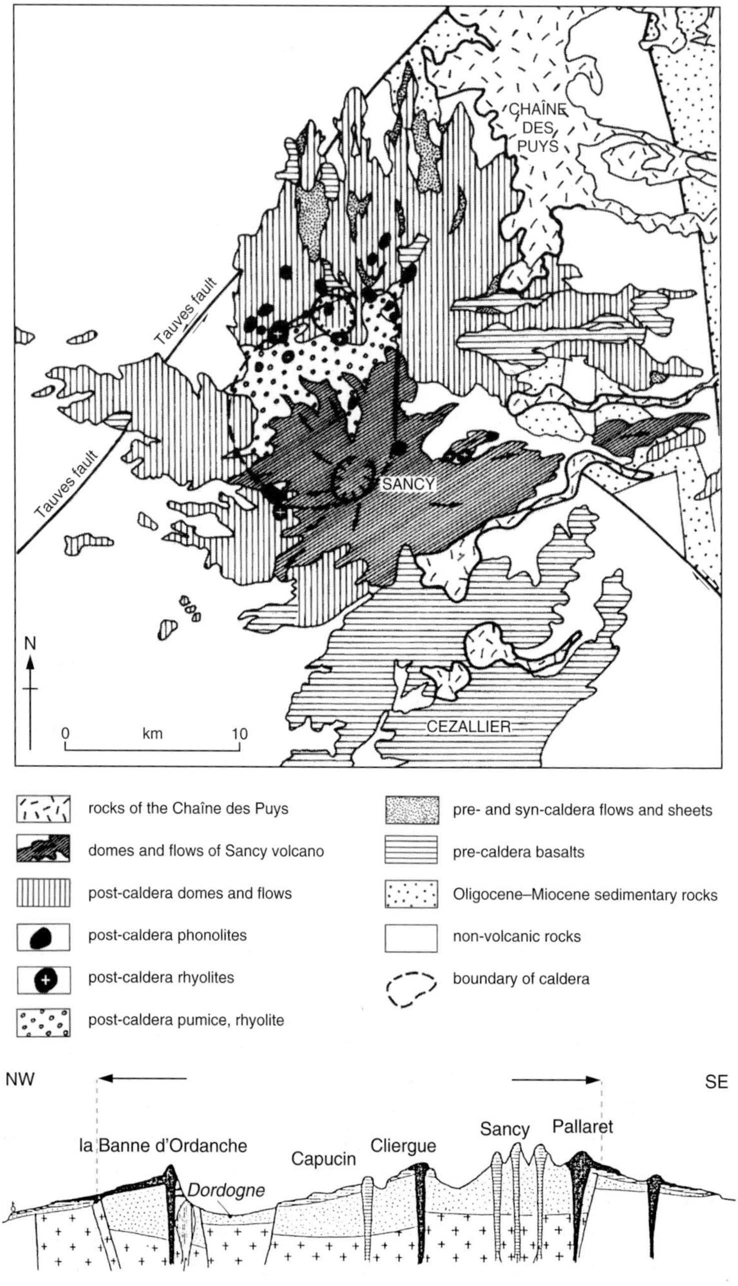
CHAÎNE DES PUYS
Tauves fault
Tauves fault
SANCY
CEZALLIER
N
0
km
10
rocks of the Chaîne des Puys
domes and flows of Sancy volcano
post-caldera domes and flows
post-caldera phonolites
post-caldera rhyolites
post-caldera pumice, rhyolite
pre- and syn-caldera flows and sheets
pre-caldera basalts
Oligocene–Miocene sedimentary rocks
non-volcanic rocks
boundary of caldera
NW
SE
la Banne d'Ordanche
Dordogne
Capucin
Cliergue
Sancy
Pallaret

Figure 6.3 (Top) Simplified geological map of the Sancy Massif (© Brousse 1960). **(Foot)** Geological cross section through the Sancy region.

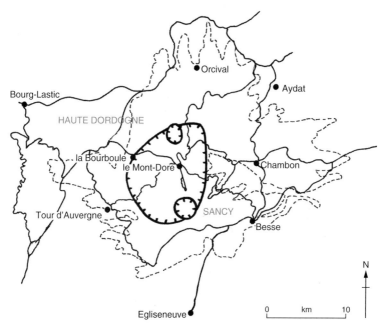

Figure 6.4 Boundary of Haute-Dordogne and Sancy caldera depressions with respect to main routes and towns.

way, with the high-temperature phases (e.g. olivine and pyroxene) settling out of the magma under the influence of gravity and accumulating near the bottom of the magma reservoir, then there is every reason to suppose that the magma body would become zoned, and the residual liquids – occupying the upper part of the reservoir – inexorably change in composition towards undersaturation. On this basis it is easy to envisage the generation of the undersaturated rock series.

This leaves the issue of how the saturated series evolved. Bearing in mind the magma collected within, and had to rise through, granitic crust before attaining the surface, the saturated series may have developed because there was assimilation of silica-rich rock by the basalt parent. Its subsequent differentiation eventually led to saturated (rhyolitic) differentiates. Although some measure of crustal assimilation may have taken place, we should note that the rock types developed within both the earlier Haute-Dordogne caldera (building the older Monts-Dore volcano) and the younger Sancy volcano, are virtually identical. This argues for the co-existence of two kinds of alkaline magma, both of which evolved by crystal fractionation of an alkali-basalt parent, eventually giving rise to the silica-oversaturated and -undersaturated series. This occurrence is supported by the presence in many places around the volcano of rocks in which many

73

quite small basic inclusions are disseminated among a more silicic host. In other words, at Sancy we find that mixed magmas and **hybridized** rocks have evolved. The simplest way of explaining such an occurrence is to suppose that these were produced when fresh influxes of basic magma entered the zoned chamber beneath the volcanic edifice, disintegrating as they did so. Because the fresh influx of magma would have built up the magma pressure once more (after the preceding eruptive cycle), it is very likely that many surface eruptions were triggered by such mixing events, which is why the hybridized rocks are found so widely in the area.

The excursions within the Massif du Sancy can be most effectively followed from centres such as Murol, le Mont-Dore or la Bourboule, but travelling from Clermont-Ferrand to the area takes a little less than two hours, so it is possible to pursue some of the itineraries from farther afield. The areas covered by the various excursions are outlined on Figure 6.3.

Excursion 1

Murol–le Marais–Tartaret–Lac Chambon (walking tour)

Introduction

Prior to the eruption of Tartaret volcano, about 12000 years ago, the Chambon River flowed off the southeastern flank of the Sancy volcano. However, when the eruption produced a large basaltic cinder cone and substantial lava flow, this blocked the exit from the valley and dammed what was to become Lac Chambon. During the course of the excursion, fine views are to be had of the lake, the Chaudefour Valley, the Sancy massif, the cinder cone of Tartaret and its associated hummocky lava flow, the modified landslip scar adjacent to the Dent du Marais, and of inverted relief in the older basalt flows upon which Chateau Murol stands and which builds the Plateau de Bessolles. The products of eruption and glaciation can both be studied at convenient exposures. The excursion is best completed on foot, but can be pieced together with the aid of a vehicle, provided there is a driver who is able to drop off the party at the beginning and meet it at the end of the various walking sections.

Starting point
Murol (see Fig. 6.5).

Itinerary
Leave the town on the D996 and walk west towards Chambon-sur-Lac.

1. The steep-sided cone of Tartaret rises on the left-hand side, with scoriaceous basalt in the roadcuts, much of it exhibiting spheroidal weathering. Just before le Marais, a small quarry comes into view on the left. Here the Tartaret basalt flow is overlain by buff-coloured glacial beds of outwash origin. At the highest level there is a very poorly sorted blocky bed, which is probably a **tillite** horizon.

2. Return to the junction between the D996 and D5. Begin the slow climb up the hill. In about 135m a wall of large lava blocks protects a quarry set back from the road on the north side. (A parking spot can be found just past the locality, on the right-hand side.) This is an interesting

75

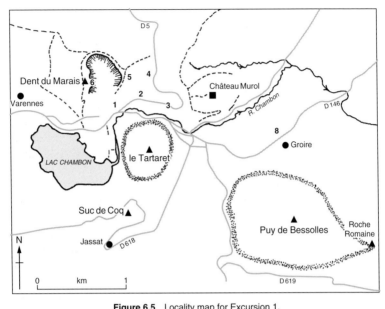

Figure 6.5 Locality map for Excursion 1.

exposure, which is gained by climbing among the blocks then scrambling up scree made from loose cinders. In the upper levels, dark Tartaret cinders are seen to lie unconformably upon glacial deposits in which there are also some cinders; there is a clear erosion surface between the two. Within the glacial beds are thin beds of dark cinders that alternate with pumice and fine ash. Bomb sags occur in places (Fig. 6.6). At road level are outcrops of buff-coloured clay-rich Oligocene sands.

3. Continue up the hill, negotiating first a sharp right-hand bend. Columnar jointed basalt is exposed on the left. A little farther around the bend there is an exposure of very weathered granite. There is then a left-hand bend in the road. Columnar jointed basalt comes in on the left-hand side, near to the water-authority building. The flow is underlain by buff-coloured ashes mixed with dark basalt and reddened lava clasts. From here there is a fine view down over Chateau Murol, perched atop its lava butte.

4. Having gone around the tump of lava, a large layby lies to the left and there are fine westerly views towards the Dent du Marais. This official parking area surrounds an information board that explains something of the local geology and flora. The thrust of the story is that the castle is built on a remnant of lava that flowed along a valley incised in Oligocene–Miocene marly muds (age 38–55 million years). Subsequent

erosion of the friable sediments has left the flow as inverted relief. A similar story applies to the Plateau de Bessolles (also visible from here). The Puy de Bessolles lavas were erupted between 19 and 10 million years ago.

5. *Saut de la Pucelle* Continue a further 150 m, whereupon a marked path takes off up the wooded hillside to the left. This gradually climbs up onto the summit of the lava plateau and reaches Saut de la Pucelle (literally "maiden's leap"), a magnificent cirque cut into the hillside, itself a part of the core of the ancient volcano of Dent du Marais. Standing on the edge of the cirque and looking towards the vertical wall of Dent du Marais, one can agree that the name is extremely apt. Various explanations have been put forward to explain this landform; some consider it to be a volcanic crater, others a landslip scar, others again a combination of the two. Whatever the truth of this matter, there is no doubt that the east and north walls of the feature are incised into compact lava that exhibits some of the strangest cooling joint patterns I have ever seen. The lower part of the cliff shows fairly regular columns, but the upper 80 m or so is a mass of curving joints that, to my mind, suggest mushroom forms. Exactly what they represent in terms of origin, I am unsure, but perhaps the upper part is either a sticky lava extruded through large tubes or is a strange laccolith.

6. *Dent du Marais* Walk across the back of the cirque, approaching the face of Dent du Marais. This spectacular wall is what is left of a volcanic **diatreme**, filled with a mixture of coarse chaotic breccia and tuff. Bear left downwards for a short distance until it is possible to stand close to the

Figure 6.6 Bedded cinders in quarry near Marais, Excursion 1, stop 3. Note the bombs, bomb sags and lamination of lower units.

precipice. Standing on its very edge is a worthwhile but scary experience (take extreme care here), which also allows inspection of the actual diatreme-filling material, a very coarse explosion breccia. (Because the wall faces east, it is best seen in the morning, when it is fully lit.)

7. Having braved the Dent, re-ascend slightly and follow a marked path that goes off to the left (west) to Croix de Barbat, passing another prominent breccia knoll on the way. At the cross, turn left, down hill, signposted to Varennes. A prominent knoll comes into view straight ahead; turn left again before it and follow a farm track that leads back to the hotels on the northeast corner of the lake (refreshment available here).

8. *Groire quarry* Walk (or drive) along the D 996 towards Murol. It is now possible to extend the walk or end the excursion here. To extend it, bear right up the hill at the first fork in Murol and bear right again on the road signposted to Groire and Sapchat (D 146). Walk along here for about 600 m, with the hummocky basalt flow from Tartaret lying on the left-hand side. Just as the road levels out, a large quarry is encountered on the left. Although most of the cone and lava that provided the material for this enterprise have now gone, there are huge piles of magnificent basalt bombs close to the road (Fig. 6.7). Some are breadcrust bombs, others have the typical spindle shape, others resemble stone papaya fruit. From the quarry floor there is a fine view towards Murol and the basalt plateaux of inverted relief. Return to Murol.

Excursion 2

Murol–Chambon-sur-Lac–Chaudefour Valley–Rocher de l'Aigle–Besse-en-Chandesse–Lac Pavin

Introduction

The peripheral deposits on the southeast side of the Massif du Monts-Dore are explored in this excursion, which ends at Lac Pavin, traditionally considered to be one of the most perfect examples of a trachytic maar anywhere in the world. However, in recent times, the suggestion has been made that it should be viewed as a crater formed by the evacuation of the magma chamber that emptied when neighbouring Puy de Montchal erupted, 6500 years ago. There are some excellent landforms to be found within this trip, and the ancient hilltown of Besse-en-Chandesse provides further variety and interest and, indeed, a good base either for an overnight stay or as a staging post for exploration of the wild lake country north of Cézallier, en route to the Cantal.

Figure 6.7 Basalt bombs in the quarry at Groire, Excursion 1, stop 8.

Starting point
Murol or Chambon-sur-Lac.

Itinerary
If starting from Murol, take the D 996 out of town towards Lac Chambon and stop at the bottle dump at Varennes, at the west end of the lake. If beginning at Chambon-sur Lac, drive in the opposite direction to the same point. There is plenty of room to park here.

1. *Quarry at Varennes* On entering the quarry, you are confronted with a bewilderingly complex pattern of cooling joints in a compact grey lava, similar in all respects to that seen in the upper part of the cirque at Dent du Marais (Excursion 1). The joint pattern indicates that at least two dykes cross the west side of the quarry face. On the east side there are exposures of basement granite.

2. Drive towards Chambon-sur-Lac along the D 996 and park near the Centre Ville sign. Walk a short distance along the main road to a steep cliff on the right-hand side. A jointed lava similar to that in (1) above is exposed in the upper part of the cliff and this overlies fresh grey granite of the basement at road level.

3. *Cascade de la Voissière* Follow Centre Ville signs, turning left off the D996, and through the lovely old village of Chambon-sur-Lac. This has a fine twelfth-century church and a fifteenth-century stone cross, which are well worth a brief visit. Take the D637 (signposted to the Chaudefour Valley) and drive through the hamlet of Voissière; in a further 500 m, stop on the right-hand side in a layby, which provides nice views of Cascade de la Voissière, which cascades over a hard volcanic capping surmounting the basement granite.

4. *Information board – Vallée de Chaudefour* Continue up the valley, with granite forming the lower ground, and mixed volcanics with a hard jointed lava cap above. On the left-hand side is morainic debris through which the road cuts in places. The U-shape valley opens out wonderfully ahead and a parking bay on the side of the road allows a stop to be made at the informative display regarding the geology and flora of the Chaudefour Valley. This has been a protected area since November 1960 and is a paradise for geologists, wildlife enthusiasts and walkers.

Geology

At the head of the valley lie the high summits of Puy Ferrand and Puy de Sancy, with Puy de Cacadogne and Puy de Crebasses (a large Strombolian cinder cone) to the right. If the lighting is good, the prominent dykes of la Rancune, l'Arche and le Moine will be seen protruding from the trees on the north side of the valley; these were intruded 350 000 years ago and were major feeders to the eruptions that built the Sancy volcano during the period 800 000–200 000 years ago. Just to the left of la Rancune are the remains of a large dome, Crête du Coq, which exploded during a Peléan eruption about 330 000 years ago. The last eruptions associated with the main part of Sancy volcano were Strombolian in type and they built both Puy de Crebasses to the north and the Montagne de la Plate to the south of the valley. These, in turn, were covered by extensive long runout flows, which now are left as benches of inverted relief at Plateau de Durbise, Chastreix-Sancy and Plaine des Moutons. A geological cross section of the structure in this region is given in Figure 6.8.

1. *Optional side excursion* It is possible to continue up the valley to the Maison de Réserve Naturelle. Just prior to this is the lovely Buron de Chaudefour, a popular restaurant/bar, with a large parking area here and informative displays. It is also a fine base for walks in the upper reaches of the valley. It is possible to walk from here up to St-Anne (a thermal area where the mineralized waters are at 18°C). This point lies right on the boundary of the Sancy caldera and it is the brecciated hydrothermally altered material along the margins of the volcano that provides easy access for the warmed rising from below.

2. *Rocher de l'Aigle*   Return to the junction between the D36 and D637 and drive up the tortuous road to Rocher de l'Aigle. This has been etched out in columnar-jointed flows left as inverted relief and it provides one of the finest vantage points for the geomorphological features of this fascinating and beautiful region. To the west is the Chaudefour Valley (Fig. 6.9), and below to the northeast lies Lac Chambon and its valley, with the lava plateaux on either side and le Tarteret at its end. One can also pick out Dent du Marais and Saut de la Pucelle, the Plateau de Bessolles and the succession of lava plateaux that stretch away towards St-Nectaire and Olloix.

Continue along the D36 towards Courbanges. Shortly before the village a sign points to Cascade du Cheix where the river falls across a

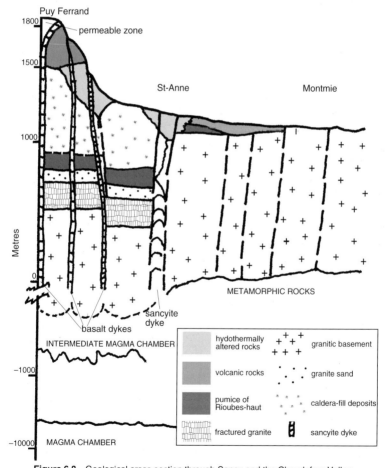

Figure 6.8   Geological cross section through Sancy and the Chaudefour Valley.

Figure 6.9a View from Rocher de l'Aigle towards Sancy.

Figure 6.9b Towards Lac Chambon. Note the glaciated profiles and well developed cirque.

step in the basalt plateau. Pass through le Verdier and descend towards the old town of Besse-en-Chandesse. This is a picturesque place with a twelfth-century church and an old quarter worthy of exploration. There are also good restaurants and hotels. Leave the town via the D149 and in about 2.5 km bear left to Lac Pavin.

3. *Lac Pavin* This crater and the neighbouring cone of Puy Montchal were the last to be formed in the region. A large parking lot (usually rather congested in high season) is to be found near the lakeside, where there is a shop and bar/restaurant. One can park here and complete a full circuit of the lake. However, I prefer to continue driving around the lake, clockwise, along a narrow road that climbs up on the rim, where there is a large parking area and a fine viewpoint with informative signboard. Follow the footpath westwards through the trees until a fork appears.

4. *Puy Montchal* Take the left path and climb up the flank of Puy de Montchal. This prominent **leucobasaltic** (hawaiitic) scoriae cone rises from the plateau on the south side of the lake to a height of 1417 m and is one of the youngest volcanoes in the region (age 6500 years). The path reaches the summit, from which there are fine views. The volcano threw out rust-coloured cinders over a wide area and extruded three flows. One poured out southwards over marshy ground, with the result that its surface is a mass of rootless cones and tumuli caused by vaporization of the underlying water during eruption. This also explains the existence of the Creux du Soucy, a lava cave 60 m long and 20 m wide. The second flow extended towards the west, reaching the Clamouze Valley, and the third spread eastwards, filling the Estivoux depression and blocking the valley glacier of Couze Pavin to the north-east. Down stream from Besse it is possible to see an old evacuated flow channel, and down stream from Cotteuge there is a very fine surface of blocky lava, clearly visible from the D978. Having taken in the view, it is possible to track down into the flat-floored crater where there are small outcrops of scoriaceous basalt, before returning to the summit and the main path.

5. *Lac Pavin* (continued) Return to the rim path by the same route and complete the circuit of the lake (without taking in Puy Montchal, this takes one hour). The lake is a traditional textbook example of a trachytic maar. The depression measures 1000×900 m and is occupied by a lake 750 m across and, at its deepest point, 92 m. Explosion breccia drapes the perimeter and, in addition to fresh trachyte (benmoreite to be precise), this contains exotic blocks of basalt, trachyandesite and basement granite. The **annulus** of breccia extends over an area measuring

15×6 km and is 15 m deep close to the crater rim. However, fine-grain tephra extends much farther and it has been recorded as finely pulverized grains in organic layers, in both Cézallier and the Cantal, reaching as far as St-Flour, 35 km to the south of the focus of eruption.

Excursion 3

Murol–Col de la Croix St-Robert–Grande Cascade–le Mont-Dore–Sancy summit

Introduction
This itinerary provides relatively easy access to the volcanic deposits on the eastern flank of the Sancy–Monts-Dore volcano, a descent through the massive sequence of the Monts-Dore caldera wall, a drive along the floor of the glaciated valley of Monts-Dore, and an ascent to the summit rocks of Puy de Sancy. You can also walk down to the starting point along the rocky shoulder that bounds the Monts-Dore Valley on its eastern side.

Starting point
Murol or Chambon-sur-Lac.

Itinerary
Leave the town along the D 996 and drive along the northern side of Lac Chambon. Beyond Chambon-sur-Lac enter the narrow valley, which has steep enclosing walls of granite capped with basalt and other volcanic rocks. In about 1 km, hairpin bends begin to appear and the road climbs up the flanks of the Sancy volcano, leaving the basement granite behind. On reaching the junction with the D 636, turn sharp left and begin to ascend tortuously up the slope towards Col de la Croix St-Robert.

1. *Col de la Croix St-Robert (1451 m)* On the way up to the col, the flattish-topped Plateau de Durbise lies to the left and the steep-sided conical mountain of Puy de l'Angle rises on the right. This is one of the parasitic massifs (mainly domes), built from doréite. Parking is available about 300 m beyond the col, the point to which one descends via the optional walking route down from Puy de Sancy at the end of the day. (In spring, the meadows hereabouts are a carpet of wild daffodils.)

2. *Descent of the Grande Cascade* The road now skirts the southern flank of Puy de Mareilh. Where the road has been widened, many blocks of vitric sancyite can be seen; these represent hot flow products from the dome of Puy de Mareilh. Beyond the col at 0.5 km a marked path leaves the road on the left-hand side; it is marked in green. Follow it and descend to the stream bed, beyond which the track follows the western

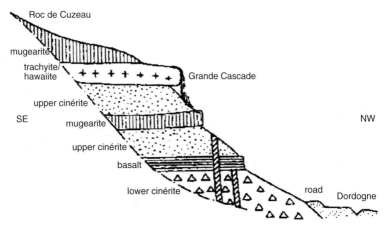

Figure 6.10 Geological section through east wall of Monts-Dore Valley, at the Grande Cascade (© Glangeaud 1969).

side of the little valley. Blocks of phonolite and breccia abound. In about 1 km the path reaches the edge of the precipice and there is a wonderful view down and across towards Puy Mary and Lac de Guéry, with the Banne d'Ordanche directly ahead. The path now plunges into the abyss, hugging the sides of the cliff and passing down the stepped lava–pyroclastic sequence depicted in Figure 6.10. The main fall plummets over a massive 0.36-million-year-old trachyte flow.

At the foot of the cliff, the track enters a wooded area and finally reaches the D 36 road again. There are now two alternatives:

- cross over the road and take another path down into le Mont-Dore (it reaches it near the Etablissement Thermal)
- continue left along the road and descend to the D 983 junction.

From either point, transport can now take the party southwards along the valley floor to the foot of the cable car station below Puy de Sancy. At the base of the ascent there are toilets, shops and a bar/restaurant. On approaching the headwall of the valley it is possible to see the cascade of the Dore River high up the cliffs, some distance to the left of the line of the cable car. It is only lower down that it joins the stream of the Dogne, to become the infant Dordogne River. The lava–pyroclastic sequence of the Durbise–Grande Cascade Cycles are well seen in the bounding walls of the Dore Valley on either side of the road.

3. *Cable car ascent to Sancy summit*   The cable car departs regularly and costs about FF30 for a return trip. (Be warned: there is a gap in the service at lunchtime.) The ascent is dramatic and very steep. The view down the Dore Valley is striking and, close to the summit station, magnificent **arêtes** and razor-sharp ridges have been etched in sancyite dykes. On

Figure 6.11 The glacially modified valley of Monts-Dore viewed from Sancy summit. Note the upstanding sancyite dyke in the foreground.

reaching the summit station there is a bar/restaurant and a wide observation platform with spectacular panoramic views (Fig. 6.11).

4. *Walking ascent to Puy de Sancy*   The cable car leaves the visitor some way below the actual summit. Ahead lies the path that ascends via several hundred steps to the summit rocks. In spring this is often covered by snow. The climb to the top takes about 20 minutes. At this point there is a table d'orientation and a geological explanation of the formation of the volcano, both very good. On a clear day it is possible to spot Mont Blanc in the distance to the east, and the Cantal summits may be seen to the south. Many well jointed sancyite dykes are prominent from this vantage point, as are the many glaciated valleys that radiate out from the focus of the massif. The summit outcrops are a mix of sancyite domes and dykes with some breccias.

The view to the south is down the U-shape Dore Valley; on the left-hand side the steep walls are crowned by Puy de Cliérgue and le Capucin; on the right are the summits of Puy de Cacadogne and Roc de Cuzeau (a sancyite dome). It is over these latter summits that the walking descent to Col de la Croix St-Robert passes.

5. *Descent options*   Three options are now available:
   - *Cable car*   Re-descend to the valley floor via the return cable car (5 minutes).
   - *Short path*   Follow the well marked path eastwards from the summit rocks. This leads to Col de la Cabane and Pan de la Grange. In front of the conical summit of the latter, the path diverges; take the left-hand fork and go down the wide valley to the foot of the cable car station (route GR4E, which takes about 25 minutes).

87

Figure 6.12 The descent route as seen from Puy Ferrand, with Puy des Crebasses (right), Roc de Cuzeau (centre) and Puy de l'Angle (left).

- *Ridge descent to Col de la Croix St-Robert*   Follow the route described above to the fork in front of Pan de la Grange, but now take the right-hand path and climb around the southwest flank of the hill and up onto the ridge. This leads to Pic Intermédiare and also to Puy de Cacadogne (1785 m). The view down on the right-hand side is very good; the beautiful glaciated Chaudefour Valley is seen with the razor-sharp dykes of Dent de la Rancune and la Crête du Coq rising out of the scree just below the ridge. Below the path itself, on the left-hand side, are a couple of spectacular jointed dykes, trachytic lavas and massive breccia outcrops belonging to the Durbise cycle. Similar rock types are encountered as the path follows the ridge crest towards Roc de Cuzeau (1737 m). It is now down hill, all the way to the col, taking about 90 minutes (Fig. 6.12).

Excursion 4a

Le Mont-Dore–la Bourboule–le Salon-du-Capucin–le Mont-Dore

Introduction
The two following itineraries can be pursued separately or merged into one continuous day out. The first allows for descent by funicular railway to le Mont-Dore from le Salon-du-Capucin, or return by car to base. The

second continues from Capucin to Tour-d'Auvergne, where the traveller can find a new base or return either to the Sancy area or press on southwards to the Cantal.

Starting point
Le Mont-Dore town centre.

Itinerary
Leave le Mont-Dore on the D 996 (towards le Queureuilh), keeping on the east side of the river and railway. The road turns west through les Marais and approaches the Petrifying Springs near Taillerie-du-Sancy.

1. *Taillerie de Sancy* Pull off into the large car-park on the left-hand side of the road. The museum/shop located close to the spring is worth a visit. It houses some fine specimens of rocks, minerals and fossils, and there are also souvenirs and books.

2. *Breccia outcrops beside the D 996* Cross the road (with care), walk eastwards, and inspect the cliffs on the north side of road, cut in a weathered feldspathic intrusion through which runs a deformed basaltic dyke.

   Leaving the car-park and driving west along the D 996, further outcrops, this time of volcanic breccia, line the roadside and continue until le Genestoux. Shortly after passing through this hamlet, a flow of well jointed basalt can be seen and is overlain by cream-coloured rocks.

3. *Quarry 0.5 km west of le Genestoux* Here are exposures (lower face) of weathered pale phonolite in the form of a dome, with low-angled columnar jointing. Above the lava, in the upper face, is what appears to be a sancyite flow.

   Continue along the D 996 and keep right at the next main fork, along the D 219. In about 100 m, pull over near where a minor road turns off to the right, signposted to l'Usclade. A large quarry face rises behind the houses.

4. *Quarry at Planches* This quarry was used for both glass-making material and abrasives. Access is restricted, but it is worth the visit because of its stratigraphic importance. The lower part of the face is seen to be cut from a chaotic rhyolite breccia overlain by a strongly laminated ignimbrite flow. This has an irregular upper contact with a dark grey trachyandesite intrusion that caps the quarry face. The importance of this outcrop is to be found in its position (close to 1000 m elevation) and its potassium/argon age (*c.* 7 million years). This places it firmly towards the base of the volcanic sequence and, as such, is an important pointer to the style of activity occurring at that time.

   It is now possible to make a detour northwards into the sequence of la Banne d'Ordanche. To do so, continue along the road to the village of Murat-le-Quaire and then strike northwards along the narrow D 609, which eventually wends its way up and around the Ordanche plateau. Most of the interesting outcrops come after the Village de Vacances.

- About 500 m beyond the turnoff to the Village de Vacances, the road intersects outcrops close to the original roof of the rhyolite dome exposed in the preceding quarry. It has a **spherulitic** fabric.
- The road now twists and turns for a while before straightening out and striking off west-northwest. Roughly halfway along the straight section, there are outcrops of weathered ordanchite, lavas associated with the basal part of the Banne d'Ordanche volcanic centre.
- Just before the road becomes too narrow for further progress by anything other than a 4×4 vehicle or horse (i.e. just before the turning circle at the base of the tele-ski), exposures of dark ankaramite come in, the rock being rich in small phenocrysts of pyroxene.

If there is not time or inclination to make this diversion, go back to the road junction between the D 219 and D 996 after visiting le Planche quarry, and follow the latter into la Bourboule. This is a good place to stop for food, drink, shopping and it also has some nice buildings. This town is located hard up against the ring fault that defines the caldera structure of Monts-Dore (the Haute-Dordogne caldera of some writers) and, indeed it is possible to discern cross-cutting boundary faults adjacent to the Etablissement Thermique "Choussy" in the centre of town. Having explored la Bourboule, take the D 888, which strikes off southwards up the hillside from the centre of town, just beside the large church. This passes through les Graffières and climbs tortuously up and around Roches de Vendeix. During this time one is within the outcrop of the varied volcano-sedimentary rocks belonging to the Cinérites Supérieures sequence.

5. *Roche de Vendeix* This prominent landmark can be ascended easily and it provides a fine panorama across the Bourboule region. Continuing up the hill beyond this rocky knoll, with its well developed columnar jointing, take a sharp left-hand turn onto the D 213 and strike off northeastwards across the northern flank of the Montagne de Bozet. About 1 km along this road there is a fine view of Rocher de l'Aigle and the mesa-like feature of Roc du Sanglier. Eventually one encounters a crossroads, at which point turn right and continue to the car-park at le Salon-du-Capucin.

6. *Le Salon-du-Capucin* This spot lies immediately below the volcanic spine of Capucin, a sancyite intrusion. Rather than climb to the summit

of this feature, walk down through the trees to the funicular station; from this 250 m-high spot there is an excellent view across the valley towards the Grande Cascade cliffs, over the town itself and across to Puy de la Tache.

It is possible to descend rapidly to le Mont-Dore from here. The funicular railway opens between 10.00 and 12.00, and from 14.30 until 16.30 each day and provides a quick means of descent. However, most visitors arrive by car or minibus and will doubtless prefer to return to base by road. To do so, retrace the route as far as the D213 then, at the crossroads, turn right, and begin the descent to le Mont-Dore.

7. *Viewpoint at height 1144 m*   In about 0.5 km the road takes a long swing around to the right. Just before it does so, there is a layby on the left-hand side. Pull off here. From this point there is an excellent view of the upper massive trachyte flows in the wall of the Monts-Dore Valley and of Roc du Sanglier.

8. Continue down the D213 for a further 0.5 km to a pullout on the left-hand side with a wayside seat. Again, there is a fine view towards Grande Cascade. Of greater interest, however, is the rock sequence exposed on the opposite side of the road: a 15 m-thick sequence of finely bedded buff and brown volcaniclastics belonging to the Cinérites Supérieures succession (Fig. 6.13). There is ample scope here to argue the relative importance of pulsatory eruptions versus fluvial deposition and, indeed, other possible interpretations.

Continue down the road and enter le Mont-Dore.

Figure 6.13 Finely banded buff and cream coloured volcaniclastics beside the D213 on the descent to le Mont-Dore.

Excursion 4b

*Le Salon-du-Capucin–Station de Chastreix-Sancy–Chastreix–la Tour-
d'Auvergne–Bort-les-Orgues*

Itinerary
From le Salon-du-Capucin, return to the crossroads and turn left along the
D213. At the junction with the D88 keep left and continue on the D213
towards the south. Eventually the D616 forks off on the left towards
Station de Chastreix-Sancy.

1. *Station de Chastreix-Sancy* If the weather is clear, it is worth taking the
   trouble to drive along this road to the ski village, which offers warmth,
   food and drinks, whatever the weather. Thus, it provides a welcome
   respite from the exigencies of travelling around. Furthermore, on the
   way up this road, and just above the turnoff to Roc de Courlande, there
   are extensive roadside outcrops of pale sancyite. Large **sanidine** crystals have weathered out and can be collected easily. The same rocks
   outcrop all the way up to the top.

2. *Chastreix* Return to the D88 and descend all the way to the isolated
   village of Chastreix. The high point here is the excellent Musée de la Vie
   Rurale, which is housed in a large building near the village centre. It is
   often necessary to collect the key. The tour here offers much local colour.

3. *La Tour-d'Auvergne* Leave Chastreix and return westwards along the
   D615 towards its junction with the D213. Here, join the latter and continue all the way to la Tour-d'Auvergne. There are good shops here, a
   very pretty church and some superb outcrops of columnar-jointed
   trachybasalt on the south side of the church. Accommodation can be
   found here and it is also possible to strike off farther west and south
   towards the Massif du Cantal, as it is a mere 15 minutes or so to the
   main route southwestwards.

4. *Cascade du Pont-de-la-Pierre* Continue along the D47 for a short distance, with granite exposures beside the road; soon there is a sign to a
   small waterfall. Park up here and take the short walk up the little valley,
   through the trees. Here the stream pours over a ledge cut in grey porphyritic granite rich in both hornblende and biotite. The rock is well
   jointed and quite fresh.

5. *Roadside lava outcrops* Continue for a further kilometre or so until, on
   a sharp left-hand bend, there are outcrops of columnar-jointed phonolite, which sits upon much more weathered **vesicular** basalt below.

   The road eventually passes through Bagnols and continues across
   farmland until it reaches the main D922 near la Pradelle. Bort-les-
   Orgues is but 7 km distant at this point.

Excursion 5

*Le Mont-Dore/Murol–Ravine d'Enfer–Clé du Lac–Col de Guéry–Roche
Sanadoire and Tuilière–Orcival–Villejacques*

Introduction
This itinerary can be started from either le Mont-Dore or Murol. If from the
former, localities 1 and 2 normally would not be visited, unless time per-
mitted at the end of the day. The excursion comprises view of sancyite and
trachyte domes, volcaniclastics accumulated within the caldera depression of Monts-Dore, several landforms produced by glaciation, and visits
to the lovely Romanesque church and town of Orcival and, for those interested, the fascinating rock shop at the small hamlet of Villejacques.

Starting points
Murol or le Mont-Dore (see notes above).

Itinerary
Leave the town along the D996 and drive along the northern side of Lac
Chambon. Beyond Chambon-sur-Lac, enter the narrow valley, which has
steep enclosing walls of granite capped with basalt and other volcanic
rocks. In about 1 km, hairpin bends begin to appear and the road climbs up
the flanks of the Sancy volcano, leaving the basement granite behind.

FROM MUROL
1. *Bressouleille to Col de la Croix Morand* The road twists and turns,
   ascending all the time. Several old rock burons can be seen within the
   pastureland hereabouts. Such rocks as are visible are from the blocky
   pyroclastic succession. Ahead lie the steep-sided hills of Puy Pouge (to
   the left) and Puy Chambon (to the right); both are domes of sancyite.
   The scree banks and gravel verges above this point are a happy hunting
   ground for collecting small sanidine crystals.
2. *Col de la Croix Morand* As you continue to climb, the prominent mountain of Puy de la Tache (1625 m) rises ahead and to the left. The latter is
   another dome of sancyite and protrudes through a dark basalt flow that
   outcrops to the right of the road. (It is possible to attain the summit of
   Puy de la Tache via the tele-ski lift, or on foot. This is an excellent viewpoint). Park at the col and take in the vista ahead. In the distance is Lac
   de Guéry with the rocky knoll of Puy Corde (phonolite dome) to its
   right. It is this parking spot that is reached if the optional walking
   descent is made from the summit of Puy de Sancy later in the day.
   Continuing down the D996; in 1 km the road swings to the left and

93

ahead can be seen the ordanchite plateau of Banne d'Ordanche. Losing height all the time, the route leads to a junction with the D983. Carry on down the D996 for a further 1.5 km, whereupon, just after a left-hand bend, an entrance leads into a large disused quarry (immediately south of the words "Prends-toi-garde" on TOP25 map 2432 ET).

3. *Quarry beside the D996.* Pull off into the large parking space in front of the quarry face (a coach can turn here). The imposing face has been cut in a massive phonolite flow, which drapes the underlying volcani-clastics visible in roadcuts during the descent along the main road. The patterns in the columnar cooling joints are complex and indicate that this is not just a simple flat-lying flow unit. There is scope for consid-erable discussion regarding the form of the body; even after a dozen visits, I am not entirely sure what I think!

Retrace the route along the D996 to the junction with the D983; take the latter, which is signposted to Lac de Guéry.

FROM LE MONT-DORE

Leave along the D996. During the first part of the route, exposures of mas-sive breccias occur along the roadside, and in the upper part of the ascent out of the town, roadside outcrops expose at least two **laharic** breccia units. Immediately after the first major right-hand bend, pull over the road into the quarry described in (3) above. The two itineraries now join.

In about 250 m there is a large layby on the left-hand side. Pull off here and cross the road to inspect the roadside cliffs. These are cut in volcani-clastics deposited within the old **caldera**.

1. *Roadside exposures beside the D983, south of Ravin d'Enfer* There is much to see here. The buff-coloured volcaniclastic rocks belong to the Middle Series sequence. They are often finely banded, sometimes chaotic and often contain large blocks (Fig. 6.14). The buff units range in grade from coarse to fine, may contain boulder trains, in which the blocks are up to 0.5 m diameter, and pebbly washouts. The coarser units are generally chaotic, and cross stratification is found in several places within the finer bands. The field evidence is for water deposition and the probable conclusion is that most of these are laharic deposits. Some intraforma-tional faults may also be seen.

Continue along the road and enter the Ravin de l'Enfer. The road falls away to the left but there are almost continuous outcrops on the right; however, do not park along this section (it is extremely danger-ous with traffic) but continue around a large left-hand bend, followed by a right-hander. At the end of the next straight, entry is barred to what was an old loop road on the right-hand side, at the point where the river crosses the route. Park on the right, just before the next left-hand bend.

Figure 6.14 Blocky laharic breccias along the roadside of the D 983.

2. *View from point where ravine crosses road* At this point the regional geological picture is revealed. The Middle Series outcrop at road level but are overlain by resilient lavas (mainly ankaramites and sancyites), which are a part of the Ordanche plateau sequence. This hard columnar-jointed capping can be seen on both sides of the main road. Blocks of the lava have fallen from the cliffs at several points and can also be inspected beside the old loop road, which leaves the main road at the stopping place. There is a good view towards the Banne d'Ordanche from this spot too.

3. *Laharic sequence along the D 983 in Ravin de l'Enfer* It is now possible to walk back along the road and inspect the volcaniclastic sequence, but take extreme care with the fast-moving traffic. The sequence is too complex to describe in detail, but, starting at the north end of the section, first you encounter fine-grain sand-grade units with large and small blocks of sancyite. Moving south, the grain size increases and there is an incoming of cinders alongside the lava blocks. There are several washouts and channel fills and, in places, regularly bedded units, often cut by minor intraformational faults. Deposition is predominantly by mudflow.

4. *Clé-du-Lac turnoff* Continue along the D 983 for a further 1 km; the narrow road to Clé-du-Lac joins the main route on the right. Park here. At this point there are fresh outcrops of buff cross-bedded mudflow deposits. In these there are some boulder trains, and somewhat rotten lava outcrops above.

5. *Lac de Guéry* About 1 km farther along the road there is the shallow lake

95

of Guéry. Never deeper than 11 m, this occupies a glacially deepened depression dammed at the downstream end by a flow of labradorite. There is a pleasant bar/restaurant here and also a small picnic site.

6. *Col de Guéry* At the northern end of the lake is the col, at an altitude of 1268 m. Pull off on the left and park up near Parc des Fleurs. A small entrance fee is charged here to explore a nice exhibit and botanical garden, which explains the botany and wildlife of this area. There are toilets and a drinks machine. Drive on a further 200 m and there is an extensive parking area from which there is a superb view of the upstanding rocks of Roches Tuilière and Sanadoire and of the Vallée de Chausse, in which the town of Orcival stands (Fig. 6.15).

7. *Information board and pullout for les Roches* Continue along the main road for a further 1.5 km, with outcrops of coarse volcaniclastics and some lavas on the right-hand side. Puy de l'Ouire rises to the right, above the treeline. Park up at the official parking spot and observe the fine view of the spine of Roche Sanadoire, which rises up steeply ahead. This is a spine of viscous trachyte with which are associated breccias; the body was emplaced during two events, dated at 2.1 and 1.9 million years ago. Perhaps the most spectacular aspect is the fine columnar jointing, which has developed in response to cooling. To the left, and separated from the former by the Vallée du Chausse, rises the equally impressive Roche Tuilière, also a trachyte dome and also beautifully jointed. The latter is a single-event intrusion (2 million years old). It is possible to walk down a steep path to the foot of Roche Sanadoire and inspect the rock and jointing at first hand. To climb to the very top, follow the descent path as described above, and then strike off right up the hillside, through the trees, whereupon a very slippery and steep path leads up left to the summit. This route is not easy; please note, the top is is extremely exposed and of very small area. However, for the more adventurous and well equipped, this could be a high point of the tour.

8. *Orcival* Continue along the D983 and take the left-hand turn along the D27, which descends through the trees and then along the floor of the Vallée du Chausse, to Orcival. This small town has many tourists, but is worth the visit to see the tranquil Romanesque church that graces its centre. There are several bar/restaurants, hotels and shops.

9. *Villejacques* The itinerary can end in Orcival or can be extended to visit Villejacques, where there is a geological specimens shop. Leave Orcival along the D27 and continue for about 4 km to the little village of Villejacques. On the left, just on entering the houses, the shop lies immediately beside the road. Local minerals and rocks are supplemented by specimens from all parts of the world, plus jewellery, books and gifts.

Figure 6.15 View towards Roches Tuilière and Sanadoire from Col de Guéry viewpoint.

Excursion 6

Le Mont-Dore–Pic de Sancy walking tour

Introduction
Several of the localities mentioned above may be linked together to form an extremely satisfying if fairly rigorous walking tour, aided and abetted by a funicular railway. This allows for a spectacular day out and a good sample of the varied volcanic geology of the Sancy massif. At a reasonable pace, the excursion would take about eight hours.

Starting point
Le Mont-Dore town centre.

Itinerary
From the town centre, make for the foot of the le Salon-du-Capucin funicular station. Take the car to the top and follow the path to the right for a short distance to take in the view over the town. Retrace the route and continue up the path to the restaurant/bar building. Now take the stony track to the left of the buildings and follow it up through the trees, which takes you to the west of the prominent neck of le Capucin (this is route GR30–GR41). Fork right about 150 m beyond the water tank, and then do a steep zig-zag ascent, passing stone burons on emerging from the forest. The path now climbs through open country, keeping close to the ridge and eventually attaining Puy de Cliergue (1691 m), composed of resilient sancyite. From this superb vantage point there is a spectacular panorama and it is possible to fit together the components of the geology: Durbise–Grande Cascade sequence in upper valley walls, U-shape Monts-Dore Valley, steep north face of Pic de Sancy with sancyite summit rocks cut by dykes, and so on. A geological cross section is shown in Figure 6.17.

The route now hugs the ridge crest and eventually turns east to Puy Redon (1781 m), but does not go to the summit, instead skirting the peak on its southern flank before arriving at the summit of Puy de Sancy (1885 m). There is no shortage of interesting rock up here, with many lava flows, breccias and intrusive masses of sancyite and dykes of similar composition to be seen.

Before climbing up to the actual summit, you may descend to the restaurant at the top of the cable car station for refreshment and a rest; however, it should be noted that the steps have to be ascended once more after the respite. From the summit follow the route described in Excursion 3 (pp. 85–88, locality 5); that is, follow the path around and down to Pan de la Grange, taking route G4 around the back of this tump and following

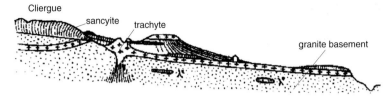

Figure 6.16 Geological cross section between the Dordogne River and le Cliergue (© de Goër de Hervé 1977).

the route down the eastern shoulder of the massif, past Pic Intermédiare, Puy de Cacadogne and up to the summit of Roc de Cuzeau (1737 m). Now it is down hill all the way across the Plateau de Durbise (where it can be boggy in damp weather) to a point just west of Col de Croix St-Robert.

Just before reaching the road, another footpath takes off to the left. For a while this seems to be going back up the mountain, but, in some 250 m or so, it turns sharply west (right) and follows the western bank of the stream that shortly plunges over the cliffs of the Grande Cascade. (This is the part of Excursion 3, localities 1 and 2.) The descent of the Grande Cascade sequence brings you down to le Mont-Dore and a well earned rest.

Chapter 7

The Monts du Cantal

The Monts du Cantal are what remains of a single very large strato-volcano, once much higher than modern Vesuvius and possible measuring 100 km across at its base. Since it reached its maximum size 3–4 million years ago, erosion by running water and ice has left the array of the peaks, ridges and valleys that we now see and among which many happy days of exploration can be spent. The geological history of this mountainous region goes back to the Palaeozoic; ancient metamorphic rocks are abundant and have yielded ages of over 400 million years; there are large bodies of granite that rose into the continental crust at much the same time. This ancient crystalline basement is widely exposed along the sides of the long valleys that radiate out from the old volcano's core and it is upon this resilient stuff that the younger volcanics were poured out. Furthermore, there are Carboniferous sedimentary rocks that formed in marginal basins ("sillon houiller") about 350 million years ago, and lagoonal sediments that were laid down in Jurassic times, about 180 million years ago.

The Cantal volcanic complex is located about 50 km south of Mont-Dore; the main peaks and ridges occupy the central part of the département, out from which fan a series of long glacially modified valleys. Within these and also on the lower slopes of the mountains, most of the main centres of population are located, which provide the best bases for exploring the massif. Bort-les-Orgues lies at the extremity and Mauriac to the west; Aurillac marks the southwest corner, and Vic-sur-Cère, Thiézac and Chaudes-Aigues lie to the south, and St-Flour at the southeast corner. Murat lies closer in on the east, along with Allanches, and Riom-ès-Montagnès is sited on the northern flank. Smaller centres, such as Salers, St-Jacques-des-Blats, Albepierre-Bredons and le Falgoux, are closer to the region's core, but the only large resort close to the peaks is Super Lioran, a wintersports complex high up near Puy Griou.

However, the region is well served by good campsites and by many chambres d'hôtes; so, should the desire prevail to stay closer to the heart of the volcano, then there are many possibilities. Because of its size, Cantal is almost impossible to tackle from a single centre, the roads twist, turn and climb for mile upon mile, and it would be too tiring to drive, walk and study geology without changing base, probably at least twice. Although Aurillac has a nice old quarter and the interesting Maison du Volcan, to my mind it is far too big and plagued by traffic to be pleasant as a major base. My own preferences include Mauriac, Polminhac, Thiézac, St-Flour, St-Jacques-des-Blats and Lanobre. Bort-les-Orgues has little to recommend it, other than its lava columns and convenient petrol stations.

General geological evolution of the Cantal region

Study of the Massif Central's rocks indicates that, if we go back some 550 million years, we have a situation where two continents were separated by an ocean, on whose floor marine sediments accumulated. Then about 540 million years ago, a subduction system was set up and this continued in operation until the two continents collided and the ocean was removed, between 400 and 350 million years ago. As is usual in such situations, the marginal sediments of the colliding continents were buried, deformed, metamorphosed and eventually raised up to form a new fold-mountain range – the Hercynian mountain chain. Over tens of millions of years the uplifted gneisses and schists were eroded, such that at the close of this first stage most of Europe was again invaded by the sea. Certainly, during Permian times, lagoons were widespread in parts of the Cantal.

The beginning of the Mesozoic saw considerable change. Towards the close of Triassic times (195 million years ago) the southwestern margin of the Cantal (around Aurillac and Arpajon) was again covered by lagoons; however, the adjacent landmass was arid, as is borne out by the widespread presence of arkoses and aeolian deposits in a belt between Figeac and Brive. Erosion inexorably wore down the continental margins to very low relief again, whereupon the sea re-advanced as the basin of Aquitaine (the ancient Mediterranean) developed. Marginal marine sediments were deposited at this time, but continued instability of the crust led to continual variations in the relative levels of land and sea, with the result that there was rhythmic sedimentation of shales, limestones and sandstone. This continued at least into early Jurassic times (*c.* 185 million years ago), when parts of the Cantal were still marine, but slowly the land rose and the sea regressed, becoming restricted to a series of large lagoons.

The Tertiary era (stage 3) began with the region being almost a peneplain and still experiencing a warm climate. However, movements in the crust and underlying mantle instigated the formation of many faults, which broke up the continental crust into a series of horsts and **graben,** in response to broad doming. The sea frequently entered the low-lying ground of the graben during Oligocene times (38–24 million years ago), forming shallow marine basins – as at St-Flour, Aurillac, Mauriac and Maurs. Many of these eventually dried up with the formation of evaporites, especially gypsum. The continued stretching of the brittle crust led eventually to eruption of basaltic lavas in the Riom-ès-Montagnès region during Miocene times (30–20 million years ago) and, about 10 million years ago, to the start of volcanic growth in the Cantal proper. This continued for a further 8 million years, presumably above a plume of hot rising mantle material. Eruptions were sometimes fissural and at other times central, and magma composition ranged from basaltic to rhyolitic, with a strong tendency to alkaline rock types. Predictably, although many lavas are found within the Cantal, there are also extensive pyroclastic rocks, including ashes, tuffs, agglomerates, pyroclastic flows and lahars.

Evolution of the Cantal volcano

The Cantal stratovolcano was built up during several major eruptive cycles; in each one the emission of lavas alternated with explosive eruptions and gave rise to a very variable suite of rock types. Its developmental history is therefore similar to that of Sancy–Monts-Dore. The Cantal eruption foci were grouped around a large area of crustal foundering, 17×20 km across. The volcanic products cover an area of 2700 km^2 and are emplaced upon a basement of Hercynian metamorphic and granitic rocks, and Oligocene sedimentary rocks. The volcano is a wide low cone, in the mountainous central zone of which the rocks are predominantly chaotic breccias buttressed by dyke complexes. The peripheral zone is more stratified, with bedded pyroclastics alternating with lavas, the oldest lavas (mainly Miocene basalts) emerging from beneath a cover of breccias at the periphery of the volcano, for instance near Aurillac, Murat and St-Flour. Together these rocks occupy a volume of around 1000 km^3, of which 80 per cent are pyroclastic in origin. Today, the highest point attains 1854 m. A simplified geological map and cross section are shown in Figure 7.1.

Construction of the volcanic edifice took place within the Upper Miocene and Lower Pliocene epochs, the sequence of events being:

1. *Infra-Cantalian basalts* (≥9 million years) These early lavas are found in the bottoms of peripheral valleys and on the margins of the massif. They are predominantly deeply eroded alkali basalts and are presumed to have been erupted from fissures.
2. *Older Cantalian latites* (8.8–8.3 million years) The first stage of cone building saw eruption of latite magma and was followed much later by alkali trachytes and rhyolites. These silicic magmas were erupted

THE MONTS DU CANTAL

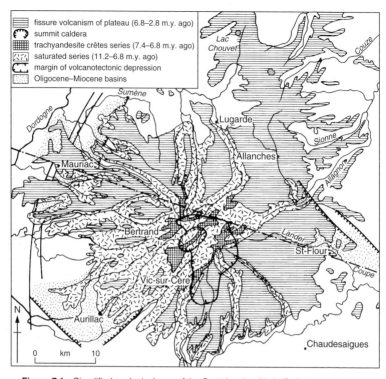

	fissure volcanism of plateau (6.8–2.8 m.y. ago)
	summit caldera
	trachyandesite crêtes series (7.4–6.8 m.y. ago)
	saturated series (11.2–6.8 m.y. ago)
	margin of volcanotectonic depression
	Oligocene–Miocene basins

Figure 7.1 Simplified geological map of the Cantal region (© de Goër de Hervé 1977).

mainly as pyroclastic flows and airfall tuffs, plus intrusion of domes and extrusion of lava domes. The first caldera formed at this stage and measured 9×6 km.

3. *Newer Cantalian trachyandesites* (8.2–7.0 million years) This stage saw the building of the second Cantal edifice. There was a wide variety of magmatic types, although the dominant lava was of saturated type and intermediate composition. Eruption was dominated by coarse pyroclastic deposits.

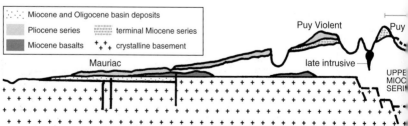

	Miocene and Oligocene basin deposits
	Pliocene series
	terminal Miocene series
	Miocene basalts
	crystalline basement

Figure 7.2 Above/opposite: cross section of the Cantal region (after de Goër de Hervé 1977).

4. *Lower breccias* Chaotic or zoned breccias and non-brecciated lava flows of large volume.

5. *Intermediate breccias* Alkaline andesites (with one or two pyroxenes) were erupted as pyroclastic flows or airfall tuffs (cinders and pumice). In the inner region the breccias tended to be chaotic, but in peripheral areas stratified agglomerates of various magmatic types were laid down.

6. *Terminal effusions* Magmas became more and more diversified, commencing within the caldera but migrating towards the exterior along radial fractures that determined the "lignes des crêtes" (i.e. "crest lines"). Trachyandesite magma was dominant, but there were also labradorites, basalts and biotite-bearing latites. The latter stages saw the emplacement of latite and trachyte domes and undersaturated lavas (**feldspathoidal**), and flows of ordanchite, plus phonolitic lava spines.

7. *Super Cantalian basalts* (≈ 6.5–4.0 million years) This was the last phase of activity (Mio-Pliocene) when vast floods of basalt were extruded, forming extensive plateaux. These covered large areas on the periphery of the massif and extended beyond its existing margins, and were accompanied by intrusion of many dykes. Emerging mainly from fissures and aligned vents, the lavas ranged from very alkaline melanocratic types (ankaramites and basanitic ankaramite with nepheline, leucite and analcime), to less alkalic types such as basalts and dolerites. A section through the sequence is shown in Figure 9.2.

Rock types and magmatic evolution

As is the case with the Sancy–Monts-Dore volcano, the rocks are divisible into two series: a saturated series (ankaramite–basalt–andesite–trachyandesite–rhyolite), and an undersaturated series (basanite–basalt; labradorite–ordanchite–phonolite). The two kinds of basalt found within the massif recall those of Monts-Dore, but are generally somewhat richer in SiO_2 than the former; however, both are undersaturated. For this reason type A magma would seem unable to produce the rock series (A) by

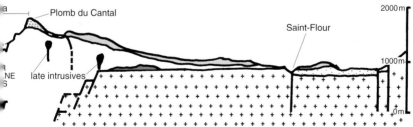

simple crystal fractionation, and several geologists have suggested that some degree of crustal assimilation must have taken place for the observed rock types to have developed by differentiation. (Analyses of a selection of Cantal rocks are presented in Appendix 4.)

Petrologists continue to argue about the degree to which the more evolved silicic magmas of the volcano owe their origin to contamination of mantle-derived basaltic magmas by fusion of silicic crust. Regardless of whether this occurred or not, and if it did to what degree, there is no doubt that some mantle-derived melts reached the surface directly, without any interaction with the crust through which they rose. These include the extensive flows of the planèze du Cantal and of Cézallier, formed during the second growth stage of the volcano. While these were being extruded, some very evolved rock types were being formed elsewhere in the area. Many of these, rather than having been the products of crustal contamination, may have been generated by such processes as gaseous transfer and crystal settling in the magma chambers underlying the edifice. Indeed, during the entire life of the Cantal volcano, the principal lava types were erupted during widely separated phases of activity. Thus, basalts tend to be scarce in trachyandesite complexes, whereas latites are very uncommon in basaltic sequences. In both cases it seems likely that incomplete mechanical mixing would have operated during their genesis; indeed, support for this notion is provided by the dispersion of < 1 cm trachyandesite relicts in the basalt groundmass of lava plugs such as the Combe de Saure, which otherwise seems to be an almost homogeneous melt. Occurrences such as this have prompted some petrologists to suggest that many or most of the intermediate Cantalian lavas may have been derived from homogenized mixing of magma types when fresh influxes of basalt entered magma chambers already filled with partially differentiated melt – something also suggested for the Monts-Dore volcano. This esoteric topic will no doubt entertain those who visit this region, take some rocks back to the laboratory and start analyzing them.

Glaciers covered the mountains in the Pleistocene; their action, combined with fluvial processes, carved out today's landscape. The most obvious pattern is the radiating valleys that fan out from the old volcano's core. During this process, many lava plugs were uncovered, flow edges etched out, and the edge of the old caldera picked out by a narrow crestline.

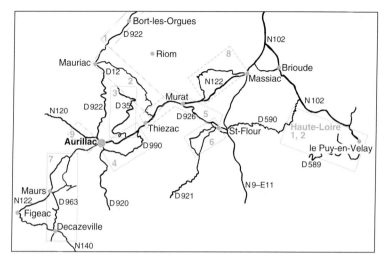

Figure 7.3   Location map for excursions in the Cantal region.

The considerable aerial extent of the Cantalian rocks necessitates being very selective in the itineraries. Those described here have been selected after visiting many individual outcrops and then linking them together in such a way as to provide you with easily accessible geology that illustrates as wide a variety as possible of rock types and geological situations. I anticipate that most visitors will also wish to imbibe as much of the magnificent scenery as is possible in a limited time frame, as well as sample the history, architecture, cuisine and culture of the region. The areas covered by each excursion are delineated in Figure 7.3.

Excursion 1
St-Avit–Herment–Bourg-Lastic–Gorges d'Avèze–Tauves–Bort-les-Orgues

Introduction
This excursion provides a useful link between either Puy-de-Dôme, or Monts-Dore and the Cantal region, and it also provides an interesting route if approaching the Cantal region direct from the north, via either Aubusson or Montluçon. It skirts the western periphery of Auvergne and lands the visitor at a spot from which to mount several days of exploration of both the basement rocks and the Cantalian volcanics. The descent into the Gorges d'Avèze is spectacular.

Starting point
St-Avit, beside the D941.

107

Itinerary

The small village of St-Avit lies 12 km west of Pontaumur beside the D 941. Its one claim to fame (in the context of this guide) is that right on the roadside is a parking area and picnic site – les Combrailles – with an interesting information board. Continue past this spot for about 3 km to le Cheval Blanc and turn right along the narrow but very pleasant D 82. This wends its way southwards across some lovely countryside.

1. *Road between Cheval Blanc and Herment* There are some excellent views of both the Massif du Monts-Dore and Puy de Dôme from along this stretch of road. Although pullouts are few and far between, there is so little traffic that this does not matter. Immediately south of the hamlet of les Pardellières, the road climbs a short hill and on the right-hand side is an exposure of weathered basement granite.

2. *Quarry 2 km south of les Pardellières* Continue for a further 2 km, the road now passing through wooded country, and on the left-hand side is a small quarry opening. This exposes reasonable outcrops of jointed basement granite and mica schist.

   The road reaches Herment and joins the D 987, passing through Lastic and reaching the more substantial settlement of Bourg-Lastic. It does not take a great stretch of imagination to realize that this was the seat of the ancient Lastic family, about whom some information is available locally. Join the N 89 in the village and turn right along it for a short distance, then left on the D 987 once again.

3. *Information board at Souilles-en-Combrailles* The road follows a particularly tortuous course for several kilometres, crosses the railway and approaches the settlement of les Gannes, where there is another information board beside the road. It then pushes on towards the edge of the plateau, takes a sharp left-hand turn and begins to descend, steeply, down the northwest wall of the Gorges d'Avèze towards the Dordogne River. Outcrops of foliated quartz–mica-schist line the road on the left; a drop yawns to the right.

4. *Bottom of the gorge – River Dordogne crossing* On reaching the lowest point, the road crosses the Dordogne River, where there is a parking spot and very pleasant picnic site. Good outcrops of quartz–mica-schist abound here, with a foliation inclined at about 40°. A track goes off towards the dam on the left, and one can walk along here to view both the river and the schist outcrops.

   The road now doubles back on itself for a while, then crosses a granite shoulder before intersecting the main D 922 road at Tauves, where there is a campsite but no hotel. Note that if you join this itinerary from the Sancy region then the route through Tour d'Auvergne meets the D 922 at Tauves. Turn right and follow the main road southwards

Figure 7.4 Columnar-jointed basalt overlying basement granite beside the D922, south of Tauves.

towards Bort-les-Orgues. The first stop (5) is on a sharp bend south of Tauves.

5. *Bend in the D 922 after valley of Mortagne*   The road climbs up towards the plateau top and suddenly goes through a vicious left-hand hairpin. It is possible to pull off right on the bend and park below the cliff. The lower part of the outcrop exposes platy basement granite, which is overlain unconformably by dark phonolite, which shows a development of columnar jointing (Fig. 7.4). This is the same flow that caps the

hillside at Bort-les-Orgues and stratigraphically lies above the andesitic breccias of the Cantal summit region and, incidentally, is believed to be equivalent in age to the ordanchites of the Monts-Dore region.

6. *Information board and pullout near la Pradelle* Continuing along the main D 922, the road climbs to the plateau top, where one or two good viewpoints are passed. Then, shortly after the hamlet of la Pradelle, there is a parking area, picnic site and excellent information board explaining aspects of the geology and ecology of the Cantal on the right-hand side of the road. (This is located about 2.5 km north of the Lanobre turnoff.)

The road now climbs up again and continues down to Bort-les-Orgues, which it reaches after a long descent. Here, there are hotels shops and petrol stations, but it is not an inspiring place to spend much time. Its main claims to fame are the famous orgues (organ pipes) and a huge dam that holds back the Dordogne, to the north. To reach the columnar-jointed phonolite of les Orgues, take a right-hand turnoff on the descent into town and follow the road up to the hydro-electric buildings and dam. It is possible to park here and take in the view. The high dam is an impressive construction, and is a good spot from which to enjoy the vista.

7. *Les Orgues of Bort-les-Orgues* Continue across the dam, whereupon the road intersects the D 979. Turn right here and continue along the southern end of the dam for about 1 km; take a sharp left-hand turn and climb up a narrow road to a table d'orientation located on top of the phonolite plateau (steps are involved). This is a very fine viewpoint that also gives access to the phonolite cliffs, with their spectacular columns. The view takes in the Dordogne Valley, the granite plateau of Artense, the Cantal and Monts-Dore. It is possible to follow a narrow path along the ridge, whence the view is even finer, with the town of Bort nestled in the valley below. Return to Bort-les-Orgues along the D 979.

If there is time, there are one or two localities close to Bort-les-Orgues that are well worth visiting, as described below.

8. *Gneiss outcrops beside the D 679* In the centre of Bort, take the narrow D 679 road which cuts up through the houses towards Champs-sur-Tarentaise. In about 500 m is a cliff of granitic gneiss on the left-hand side. It is possible to park on the opposite side of the road, on a curve in the road.

9. *Gorges de la Dordogne Information area.* A little farther along the same road is a large pullout on the right-hand side, in which stands a huge granitic gneiss monolith. Within this is very fine **augen** structure. There is a nature trail here, together with a very informative display board

that describes the Dordogne River gorge and its formation, and so on. A worthwhile stop, the outlook from this old quarry offers fine views of the lava-capped plateau top across the valley.

10. *Quarry beside the D 679*   Carry on for a further 2 km or so, until a small quarry comes in on the left-hand side of the road. Park on the right side of the road. This small exposure shows a sequence of alternating siliceous and pelitic schists. Mica is abundant and there are some quite large quartz lenses to be seen. The foliation is at a very low angle. Similar rocks outcrop all the way along the road to Champs-sur-Tarentaise. Return to Bort or, of course, one can now drive all the way through the lake country to Besse.

Note
There is an excellent campsite just north of Bort, which has direct access to the lake formed by the Dordogne barrage. Take the D 922 north and turn off left in about 3 km. This site, Municipal de la Siauve (listed in the Michelin camping guide under Lanobre), provides a very good base. Furthermore, nearby in Lanobre village, which is reached by continuing northwards along the D 922 for a further 1.5 km, then turning right, there is an old hotel in the village square, which serves excellent food at very reasonable prices. There are few frills but likely to be few complaints with the fixed-price menu.

Excursion 2
Bort-les-Orgues–Route du Falgoux–Puy Mary–Roche Noire–Vallée de Mandailles–Col du Pertus–St-Jacques des Blats–Thiézac

Introduction
This full-day itinerary begins on the periphery of the Cantal volcano, then turns along one of the beautiful radial valleys and penetrates right to the heart of the massif. The climb to the summit is spectacular, as is the descent southwards to the Vallée de Mandailles. In so doing one can study the basement rocks, peripheral basalts, phonolites, the pyroclastic sequence of the main eruptive cycles, and the late-stage saturated latite sequence. The final section is also very beautiful and brings the explorer to the pleasant old town of Thiézac, where there are several good hotels, shops and campsites.

Starting point
This excursion can be started either at Bort-les-Orgues or Mauriac.

111

Itinerary

From Bort-les-Orgues, drive southwards along the main D922 road. South of town the jointed lava capping comes into view on the right-hand side. On the left-hand side of the road are outcrops of basement mica schist. These continue all the way to the River Sumaine.

1. *Greywacke outcrops near "Ferme Auberge", Bassignac* The railway crosses the road just before Ydes, then remains on the right-hand side for a further 8km or so. About 0.5km beyond Bassignac, outcrops of steeply inclined sedimentary rocks emerge among the trees on the left-hand side of the road (take care parking here, as there is no obvious pullout). The sequence includes some coal-bearing bands, together with medium-grade greywackes which alternate with shales. The rocks are tilted at 70° and distorted (Fig. 7.5). Although a small outcrop, it is one of the best within the Carboniferous sequence that accumulated along the narrow fault-bounded basin that runs along the western edge of the Cantal.

The road continues southwards to Mauriac, and occasional sketchy outcrops of granite are to be seen along the roadside, most not justifying a stop. Mauriac is sited at the edge of an extensive plateau of basalt and is a market town of some substance and also a centre for glove making. It has three hotels and therefore can act as a base for this quarter of the region. The church in the main square, Notre-Dame des Miracles, was built during the twelfth century and is the largest Romanesque

Figure 7.5 Steeply inclined Carboniferous greywackes and shales alongside the D922 near Bassignac.

structure in this area; however, it is relatively plain inside and compares unfavourably with the churches of Orcival, St-Nectaire and Clermont-Ferrand. It houses a black Virgin Mary. Our route now lies a short distance back along the D922.

If a visit to Mauriac is not to be made, then turn off the main D922 about 1.5 km north of town, near le Vigean, and take the D678, eastwards. If Mauriac has been explored, then return northwards up the D922 and follow the above instructions. In 7.5 km the road joins another, the D12, and becomes the Route du Falgoux. We will follow this delightful road all the way to Puy Mary.

2. *Roadside outcrops near the Mars River* Immediately after joining the D12, schistose rocks outcrop on the left-hand side. The foliation is virtually horizontal. The view from here (and farther along the road) shows quite clearly the depth of the incision of the plateau by the river. There are frequent glimpses of the lava capping at the top of the plateau; the intervening slopes are built of volcanic breccias.

3. *Les Aldières* There are several beautiful old houses here, with steeply pitched volcanic "slate" roofs. The view up to the left clearly shows the columnar-jointed lava capping the basement granite. Soon this weathered granite appears at road level.

4. *Gorges de St-Vincent* The valley sides now steepen and the road enters the Gorges de St-Vincent. The prominent cliff on the left-hand side, behind the cemetery in St-Vincent, is capped by several well jointed basaltic flows; again, these lie above the bluffs of granite, much clothed by forest.

The road wends its way pleasantly along the valley floor, with the impressive walls rising to a height of 1100 m. On reaching this village of le Faulmier, there is a lovely view back down the route; ahead there are fine views up towards Puy de la Tourte. At le Falgoux itself there is a rural campsite in a very tranquil setting but with modest facilities.

5. *La Franconèche* On the left-hand side of the road are some well formed organ pipes in the basalt capping flows. There is also a fine view towards Roc du Merle, on the opposing side of the river. On the sharp left-hand bend just beyond the hamlet, the roadside granite outcrops give way to breccia containing angular blocks of sanidine-trachyte and andesite up to 1 m across (Fig. 7.6). At last the route has penetrated the saturated series of the central part of the volcanic complex that previously has lain well above road level.

As the road climbs, ahead can be seen the impressive head of the valley – the Cirque du Falgoux – with its glacially oversteepened walls. In about 2 km the road turns sharply right, at le Pont des Eaux. From here there is a fine view back down the valley with a restored stone

Figure 7.6 Volcanic breccias with large sanidine-trachyte clasts, la Franconèche, Vallée du Falgoux.

buron in the foreground. At the junction with the D680, turn right and drive for about 1 km to the foot of the impressive lava cliff of Roc du Merle.

6. *Roc du Merle* There is a layby immediately below the scree at the foot of this prominent landmark. A wonderful panoramic view down the Vallée du Falgoux greets the eye. The rock itself is built from fine-grain

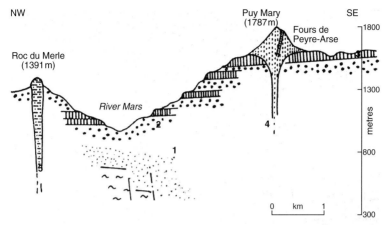

NW   Puy Mary   SE
                                                       (1787m)
                                                              Fours de   1800
                                                              Peyre-Arse
Roc du Merle
(1391m)

                                                                                1300

        River Mars
                                                                        metres

                                                                                800

                                               0   km   1

                                                                                300

Figure 7.7   Cross section through the Vallée du Falgoux (© Bout & Brousse 1969).

olivine-bearing basalt and there are some very complex cooling-joint patterns to be seen in the cliffs above (beware of falling rocks). A cross section through the Vallée du Falgoux is shown in Figure 7.7.

Retrace the route to the junction with the D680 and climb towards Pas de Peyrol. As the road passes beneath Roche Taillarde, outcrops of breccia can be seen beside the road, but it is too dangerous to stop. Continue upwards, whereupon the route steepens somewhat alarmingly and takes a sharp hairpin bend before reaching the col. Before the hairpin, nice outcrops of pinkish trachyte are found on the right-hand side. The views are magnificent.

7. *Pas de Peyrol*   There is a very large parking area here, but, be warned, it can be full in high season. There is much interesting rock to be seen alongside the road, much of the material between the parking lot and the preceding hairpin being volcaniclastic, with angular blocks of pale latite set in a pale **matrix**. This is a pyroclastic flow deposit that undoubtedly originated as the summit dome of Puy Mary was emplaced. Immediately ahead rises the summit of Puy Mary, at an altitude of 1783m. One reaches the summit via a good but steep path with many steps. The ascent takes 25 minutes and there is a table d'orientation to mark the spot. Puy Mary itself is a mass of **porphyritic** latitic andesite and fresh, well jointed, outcrops line the ascent path. The phenocrysts are of **oligoclase** and brown **hornblende**, the latter often having a flow orientation. A prominent dyke is encountered on the west side of the path. The vista from the top is quite breathtaking; in particular, the pattern of radiating glaciated valleys and intervening ridges impresses the eye. The northerly panorama sweeps west from the peak of Puy Violent, takes in the narrow ridge of Roc des Ombres,

115

Figure 7.8 Puy Mary, the summit cross. The upstanding phonolite spine of Hozières lies immediately to the right of the cross with the thick lava remnant of Roche Taillarde beyond and to the left. To the right is the deep Vallée du Falgoux.

then Roche Taillarde and Hozières, before revealing the Vallée du Falgoux and the ridge of Puy de la Tourte and Rocher de l'Aygue, which separates it from the valley of la Véronne (Fig. 7.8). The southerly view takes in Puy Griou, then the broad Vallée de Mandailles, bounded on its northwestern side by the ridge containing Puy Chavaroche and Puy de Bassierou. It is this valley that provides the descent route to Thiézac.

Although it is possible to follow the path that crowns Puy de la Tourte and Rocher de l'Aygue ridge (a magnificent ridge walk), time precludes this and, to follow the excursion, return to the col. Now continue along the D 17, southwards, towards Mandailles. The road hugs the crest that marks the top of the Cirque du Falgoux and, in about 1 km, passes the upstanding knob of Roche Noire.

8. *Roche Noire* This prominent rock is composed of andesitic breccia, which is in contact with a porphyritic trachyandesite lava characterized by a distinct platy jointing. Steeply inclined dykes cut the complex. The view down the valley head towards the phonolite spine of Hozières is particularly good from here.

9. *Col de Redondet* Immediately around a sharp left-hand hairpin, there is a parking area. To the west is the upstanding hill of la Chapeloune, cut by a series of phonolite dykes. Beyond the bend, outcrops of dark trachyandesite occur at road level, as do exposures of a much lighter trachytic type. To the southeast, the view towards the phonolitic spine

of Puy Griou is particularly fine. *Option*: It is possible to ascend la Chapeloune (it takes about an hour for the return trip) and view the summit hornblende-andesite flow, which is underlain by both breccia and trachyandesite and cut by phonolite dykes. The view of the glaciated topography is very fine from this point.

10. *Viewpoint for Roche Noire and bedded-series cliffs*   Continue for a further 300 m, where a viewpoint layby is encountered; park here. Coarse volcanic breccias outcrop along this section, and build the stratified cliffs above the road (Fig. 7.9). The view is impressive. Further down the road, a sequence of agglomerates and trachyandesite lavas comes in on the left-hand side. These belong to the intermediate breccia succession and include both airfall and ashflow deposits.

Continue down the road towards Mandailles with many coarse breccia and lava outcrops above and beside the road.

11. *Information board: Puy Griou group*   Eventually, as the road swings around through the trees, there is an information board that describes the nearby spine of Puy Griou.

Descend along the D17 into Mandailles village. This is a lovely old place and has good beer at the Hôtel au Bout du Monde ("World's End"). The village is pleasantly situated beside the River Jordanne. Cross the river bridge and, in about 300 m, turn sharp left up the hill towards Col du Pertus, along the D317.

12. *Information board: Vallée de Mandailles*   Footpaths are signed to Mandailles and Col du Pertus from this point and there is an old buron on the left-hand side. From this spot there is a superb panoramic view of the central part of the volcano, with the stratified lava–breccia sequence

Figure 7.9   View towards Roche Noire (left) and the cliffs of bedded pyroclastics north of the D17 at locality 10.

117

to either side of Puy Mary (Fig. 7.10). The vista spans the section from Cabrespine in the west, through Puy Mary, to Puy de Peyre-Arse, in the east.

The road continues to climb and shortly a well jointed dyke appears in a roadcut on the right. It is about 5 m thick and was intruded into what is now deeply weathered volcaniclastic material. At Col du Pertus there is another geological information board and layby. The road now descends towards St-Jacques-des-Blats.

13. *Exposures near Fonjouque*   On the descent, several exposures of blocky pyroclastics are to be seen. They continue to appear and, close to the burons about 1.2 km farther down the road, are some fresh and extensive exposures of unwelded pyroclastic flow deposits (Fig. 7.11).

The road eventually joins the N 122 at St-Jacques-des-Blats. Turn right along the N 122 and drive southwards. All along the right-hand side of the route are large outcrops of massive breccia (probably unwelded ignimbrite) belonging to the Older Cantalian series (latite breccias and pyroclastic flows, dated at around 8 million years ago). It is easier to inspect these nearer to Thiézac, where there are some small pullouts beside this busy road. The road to the town leaves the main route on the right. It makes a pleasant base for exploration of the southern Cantal, and has several good hotels, a campsite and shops.

Figure 7.10   Panorama of the Cantal volcano from between Mandailles and Col du Pertus.

Figure 7.11 Exposure of unwelded pyroclastic flow below Fontouquet, beside the D317.

Excursion 3
Thiézac–Polminhac–St-Simon–Col de Breul–Col Legal–Bastide–le Fau–
Fontanges–Salers–Mauriac

Introduction
The route crosses the southwest sector of the Cantal, taking a tortuous path across the Route-des-Crêtes, passing through the ancient towns of Fontanges and Salers before emerging on the main north–south road at Mauriac.

Starting point
The excursion can be started at Thiézac, Polminhac or Aurillac. The first point of reference is the village of Polminhac on the main N 122 road.

Itinerary
About 1 km west of the village, turn north along the D 158, climbing up towards Murat Lagasse.

1. *View over Cère valley from Murat Lagasse*   From this point there is a fine view over the valley of the Cère. The road eventually doubles back on itself and follows the top of the lava plateau. At both Costes and Fraissehaut there are superb views of the plateau in all directions.

    Join the D 58 near Boussac and follow the road down towards St-Simon. Just before dropping down into the town there are superb views down into the Jordanne Valley, with the Cantal massif beyond. The narrow road intersects the D 17 just over the Jordanne River. Turn left and then almost immediately right up the D 58 (signposted to Marmanhac). Upon reaching the Route-des-Crêtes (D 35), turn right. In about 4.5 km, turn left along a narrow lane towards Alquier.

2. *Table d'orientation at Alquier*   Park near the farm buildings, where there is a small quarry in trachytic breccia. Follow the marked path up to the top of the hill (997 m high), which is carved out of a well jointed flow of **aphyric** black basalt. On the way up are outcrops of a pinkish breccia with angular black clasts. The wide vista takes in Col de Peyre-Arse, the Cantal Massif and the Jordanne Valley.

    Retrace the route back to the D 35 and continue eastwards along it to Croix de Cheulles. Extensive views greet the eye as this point is approached. Go ahead along the D 35 to the hamlet of Houade.

3. *Quarry at Houade – outcrops at Col de Breul*   At Houade there is a small quarry on the right-hand side. It exposes a **polymict** matrix-supported breccia containing small black glass fragments and quartz clasts. The outcrop of this unit continues along the roadside towards Col de Breul, where it is possible to spot occasional **fiammé** within the outcrops, indicating it to be an ignimbrite flow.

Drive on to Col de Legal and descend for a further 0.25 km to a point where a rough track leaves the road on the left-hand side. Park here.

4. *Quarry just beyond Col de Legal*   A reddish breccia contains quartzose **inclusions**, lava blocks and feldspar crystals. It may be a continuation of the pyroclastic flow unit seen earlier or may be a **lahar**.

The road now reaches Col St-Georges, where there is a picnic site, parking lot and excellent viewpoint. At the col, take the narrow D 42 road to la Bastide and park beside the road in the hamlet.

5. *Mineral spring and ignimbrite at la Bastide*   A marked path descends a short way towards the river. A somewhat pathetic-looking trickle emerges from the base of a damp cliff; the latter is of more interest as it is composed of unwelded ignimbrite. While here, it is worth walking to the nearby bridge and viewing the river.

6. *Outcrops beyond le Fau*   Continue along the same road to le Fau. At the far end of the village is a splendid view over the valley of the Aspre River. Farther on again, about 1 km, and it is possible to pull off the road to inspect outcrops of compact pyroxene-rich crystal tuff. This could be more of the ignimbrite, as the matrix appears pretty well indurated.

Eventually the road begins a tortuous and steep descent towards the Aspre Valley and the D 36 road, which runs through the quaint old village of Fontanges. Roadcuts along the descent expose a laharic breccia with yellowish matrix and a poor stratification in places. Farther down, at the hairpin bend, the breccias contain even larger blocks.

7. *Fontanges – Chapelle du Monolithe*   On reaching the D 36, go straight across the road and drive to the strange Chapelle du Monolithe, which is built into a huge upstanding remnant of volcanic breccia. This structure is a real oddity; the rock from which it is constructed is a beautiful massive breccia (Fig. 7.12) that appears to represent an old neck. It is possible to climb to the summit of the neck for a view over the Aspre Valley and the village. The village itself, which is close by, makes a picturesque spot for a picnic; there is also a bar and camping site.

8. *Salers*. Leaving Fontanges, continue along the D 35 to the lovely old town of Salers, where there is a large parking area with public toilets. This town is one of Auvergne's showpieces, being sited on a lava plateau high above the River Maronne. Although it has seen some tourist development, it has escaped the worst excesses. There are some wonderful fifteenth- and sixteenth-century houses within its walls but, interestingly, the church was built outside the ramparts. The natural focus is the Grande Place, where most streets lead. At the edge of the town (literally) is the Promenade de Barrouze, a pleasant viewpoint over the river and towards the Cantal mountains. There are some chic shops and three expensive hotels. Salers is worth at least an hour of

121

Figure 7.12 Chapelle-de-Monolithe at Fontanges, a shrine built into a grotto within volcanic neck. The stone from which it is built is a volcanic breccia with multi-coloured lava clasts.

anyone's time. Leaving Salers, first take the D22, then turn off left towards St-Bonnet-de-Salers. Pass through this village and Chastenac, join the D29 and reach the main D922 road. Here turn right (north) and drive for about 2 km. The road first crosses the railway and then the river, just beyond which it is possible to park beyond the sign "L'Auze". Cross the road (with care) and join a path beside the Bar du Cascade.

9. *Cascades de l'Auze*   Follow the path briefly, until the edge of the gorge is reached and there is an impressive view of the rushing cascade. The water floods over an edge cut into a Miocene basalt flow that overlies Oligocene mudstones with some gypsum beds. Being softer, the latter have become undercut. It is possible to descend along the path that runs behind the waterfall. It takes about ten minutes to reach this point.

Return to the main road and continue to Mauriac or Bort-les-Orgues.

Excursion 4

Thiézac–Arpajon–Carbonet–Roches-du-Carlat–Rocher-des-Perdus–Col de Curebourse–Thiézac

Introduction
This is a varied and interesting itinerary, which investigates the geology along the southern edge of the Cantal. The first stop explores an excellent section through Oligocene calcareous sediments near Arpajon, fossiliferous deposits of the Aurillac Basin, which developed within a downfaulted trough on the margin of Auvergne in Tertiary times. Then, in turn, the basement schists, peripheral basalts of Carlat, chaotic breccias of the central volcanic complex and lahars of Col de Curebourse are encountered before returning to Thiézac.

Starting point
The excursion is described as starting from Thiézac, but Aurillac or Polminhac also provide suitable bases.

Itinerary
Join the main N122 road and drive to Aurillac. Having passed under the railway, take a left turn signposted D920 to Montsalvy and Rodez. Enter Arpajon and watch out for the D58 on the left-hand side; the road is signposted to Vic-sur-Cère. In a short while turn left again, along a minor road signed Cité de Puy Griou. Cross the bridge and the hill of Puy de Vaurs rises straight ahead. Park on the left-hand side, at the point where a gravel

123

Figure 7.13 Quarry at Puy de Vaurs, exposing banded Oligocene marls with hard layers of siliceous nodules.

track joins the road (Rue de Four à Chaux) as it turns a right-hand corner.

1. *Quarry at Puy de Vaurs* Follow the track up the hillside until a quarry level is reached on the right-hand side. The white, well bedded, calcareous marls provide a classic section through the Aurillac Basin sediments (Fig. 7.13). Many blocks lie on the quarry floor and it is possible to retrieve fossil specimens (*Potamides*, a gastropod). The **siliceous** bands are frequently regularly banded and make good specimens.

   Return to the road, rejoin the D 58 and drive to Carbonnat. Now turn right and in a short while join the D 990; turn left towards Vezac, then, in 0.5 km, pull off on the left-hand side.

2. *Quarry beside the D 990 near Monteydou* There has been some recent excavation here and a good clean section is provided in the foliated grey mica schists of the basement complex. There are further outcrops about 300 m farther along the same road, also on the left-hand side.

   Continue along the D 990 until the turning for St-Etienne-de-Carlat.

3. *Roadcuts at turning for St-Etienne-de-Carlat* Fresh roadcuts reveal a sequence of red cindery beds, which represent peripheral basaltic deposits of the Cantal volcano.

   Continue along the D 990 to the village of Carlat, then turn steeply up to the left (signposted "Information – le Rochers") and a good car park. The flat-topped plateau remnant of Rocher de Carlat rises immediately above the houses.

4. *Rocher de Carlat* The table-top lava feature represents inverted relief.

A series of valleys that had been cut into the Oligocene landscape were invaded about 5.5 million years ago by fluid Cantalian basalts. Subsequent erosion stripped off the sediments and basement rocks, leaving the resilient flows as mesa-like remnants with a long but narrow outcrop. This particular flow has a length of about 30 km and fossilizes an older course of the River Embène. Walk up through the car-park and follow the path up the hill. The ascent of the Queen's Staircase provides a fine view over the Cantal hills. The basalt is seen to be well jointed and the panorama from the top of the "table" is excellent. Once there was a castle here, but this has long since been destroyed. The breach that separates the two parts of the rock was once believed to have been made by the early inhabitants of this region.

Return to the car-park and rejoin the D 990, driving eastwards along it. Pass through Lessenat and Montat.

5. *Roadside cliffs on left-hand side of the D 990*   Between the Montat and the junction with the D 157 are some prominent cliffs composed of massive volcanic breccia in which angular lava blocks are set in a whitish matrix. The matrix-supported nature of this unit suggests that it may be a lahar.

Turn off left along the D 157 to Baidalhac. This is a beautiful old village from which there are excellent views of the accordant summits of the dissected basalt plateau of Cézallier. Beyond the village, turn right along the D 57 towards Jou-sous-Monjou. About 1 km farther along the road is a small wayside cross on the left, beside which is a track that leads to the plateau top and the basalt flows from which the rock is built. On a good day this makes a pleasant walk.

6. *Breccia outcrops beside the D 157 on descent from plateau*   The road begins to drop down from the plateau and shortly crags on the left expose more matrix-supported massive breccias. These contain angular clasts of basalt and phonolite up to 0.5 m in diameter.

The next hamlet is Vixe, beyond which are fine views down into the Vallée du Goul. Enter Jou-sous-Monjou and turn left along the D 54 road, which skirts a steep hillside on the left-hand side. The hill is a dome of massive phonolite. The road now passes below Rocher des Pertus, composed of olivine basalt. At the point where a small road to Fraysse leaves on the right-hand side is an exposure of vesicular trachyte, sometimes **amygdaloidal**.

7. *Col de Curebourse*   Park at l'Auberge des Monts. To the left is the glaciated valley of the Cère, the view being punctuated by small domes and spines mainly of trachyte. Ahead are several hanging valleys, indicative of past glacial activity. If weather conditions spoil the fine views that can be had from here, there are excellent exposures of laharic breccias ahead.

Figure 7.14   Detail of laharic breccia in roadside below Col de Curebourse.

8. *Laharic breccias below Col de Curebourse*   Continue ahead along the D 59 on foot, descending towards the River Cère. In about 200 m some excellent outcrops of laharic breccia are found on the right. Angular blocks of trachyte and darker phonolite up to 0.3 m across are set in a brownish matrix, which may be weakly stratified (Fig. 7.14). In places the blocks appear to be aligned and elsewhere the phonolite clasts are reddened.

9. *Pas de Murgode*   The road continues to descend to the pass, whereupon extensive outcrops of laharic breccia come in on the right-hand side (netted over in places).

Continue down to the main N 122 road and turn right towards Thiézac.

Excursion 5
Thiézac–Murat–Falls of le Saillant–Talizat–Andelat–St-Flour

Introduction
This excursion provides a link between Thiézac and St-Flour, on the south and southeast sides of the Cantal respectively and takes in some good varied geology. The traverse includes visits to a wide variety of outcrops, including the chaotic breccias of the central Cantal complex between Thiézac and Lioran, superb views over the Cantal mountains and valleys

from the summit above Lioran, the aligned volcanic necks of Bredons (near Murat), the spectacular basalt gorge and waterfall at le Saillant, a lahar at Andelat, major cinder cone at Puy de Talizat, olivine nodules at Cussac, and the organ pipe cliffs at St-Flour.

Starting point
Thiézac.

Itinerary
Leave Thiézac and join the main N 122 road, turning east towards Lioran. Drive for about 2 km and stop beside the road (take care, it is a busy route).

1. *Massive breccia outcrops, east of Thiézac* Massive, chaotic, matrix-supported volcanic breccias line the roadside. Large blocks have been dumped on the right-hand side of the route, and some good specimens can be inspected here. These and those farther along the route belong to the series of intermediate breccias of the newer Cantalian trachyandesites. Similar rocks can be found for many miles along this road, but it is usually dangerous to stop.

   Continue along the N 122, passing through St-Jacques-des-Blats, with the valley of the Cère always on the right-hand side. Just beyond les Chazes, leave the main road (if you go through the tunnel, you have missed the turning), rather taking the old D 67 road that wends its way steeply and picturesquely up on the left to Super Lioran. In about 1.5 km pull over on the right to a parking spot at Font-de-Cère.

2. *Font-de-Cère* There is an information board here that focuses on the agriculture of the Cère Valley and also provides a magnificent viewpoint over it.

   Carry on up the hill until eventually reaching the resort village of Lioran. There are few decent outcrops along this section; however, the scenery is pleasant. The road levels out and runs through a steep-sided wooded valley – the Gorge de l'Alagnon. Farther east the valley broadens and the views open up nicely to the north. Pass the turnoff to Laveissière, and the hamlet of Fraisse-Bas and, just before the turning to Chambeuil, come to fresh-looking cliffs on the right-hand side of the route. Park here in a large pullout.

3. *Basanite outcrops near Chambeuil, beside the N 122, Alagnon Valley* The cliffs are about 20 m high and comprise a thick lower unit, which weathers to a reddish colour and is very compact, overlain by a thinner, softer red band, which has a hard grey cap. Above the latter, forming the top of the cliff, is a greyish unit of indeterminate material. The lower unit is very compact and contains hornblende and biotite **microphenocrysts** set in a reddish **groundmass**. The rock is well jointed. It also

127

contains angular enclaves of quartzose material up to 150 mm long. It is a basanitic flow of Miocene age. The overlying reddish layer is apparently a weathered breccia, but cannot be reached for a close inspection.

4. *Breccias at Anterroches*   A further 0.5 km along the main road, the steep-sided hill of Anterroches rises on the left. Composed of chaotic breccia, often quite well indurated, it overlies the Miocene basanites seen earlier. The deposit may be the product of a late-stage explosion crater.

The road now enters the pleasant town of Murat, located in the valley of the Alagnon between the Miocene basanite cliffs of Rocher de Bonnevie to the north and that of Bredons to the south. In fact, it is sited at the confluence of several converging valleys once occupied by valley glaciers during the last ice age. It is an old town and has many original houses still roofed with the volcanic stone lauzes, which are cut with one semi-circular edge, so that they look like huge fish scales when hung. It is regrettable that it is overlooked by a particularly hideous nineteenth-century white statue of the Blessed Virgin, located on top of Rocher de Bonnevie. To my mind, its monstrosity is surpassed only by the red version that crowns the rock at le Puy-en-Velay. This aside, the town has reasonable shops and restaurants, is a centre for some of the best angling in Auvergne, and is a setting-off point for the aligned vents of Albepierre–Bredons, which will now be described.

GEOLOGY OF THE BREDONS NECKS

The necks are aligned along a fissure and are basanitic, containing **modal nepheline** in the case of the Bredons neck and **normative nepheline** in the case of Bonnevie. They represent the products of a Miocene event and are accompanied by explosion breccias related to maar formation. The information board beside the main D 926 road has a well designed explanation of the geology and brief history of the area (see stop 9 and Fig. 7.15).

5. *Chastel-sur-Murat volcanic neck*   Join the D 680 road on the east side of the prominent Rocher de Bonnevie and drive northwest towards l'Her-itier. The road traverses poorly exposed breccias overlain by a flow of labradorite lava. Beyond this village, after 1.5 km turn sharp right along the D 139 to Chastel-sur-Murat and park at the entrance to the village. A steep but short path climbs up to the summit of the neck. The rock is a melanocratic porphyritic basalt, in which can be seen phenocrysts of olivine and augite. It shows a development of low-angle columnar jointing. From the summit there is a fine panorama: westwards the high summits of the Cantal can be seen, to the northwest is the basalt flow of Lapsou, and the white excavations of the Faufouilloux diatomites are prominent to the east. The line of vents – the rocks of Bonnevie and Bredons – extend towards the southeast. In the far distance is the

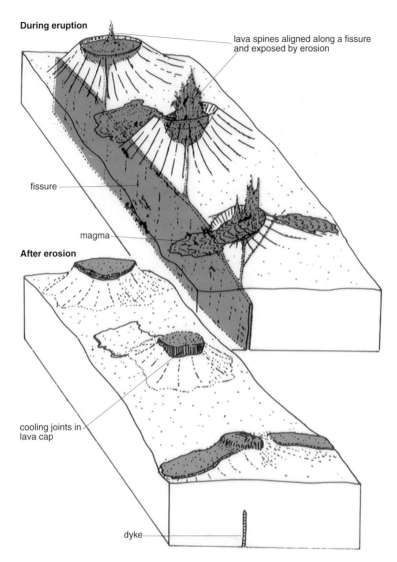

During eruption

lava spines aligned along a fissure and exposed by erosion

fissure

magma

After erosion

cooling joints in lava cap

dyke

Figure 7.15 Panorama across the region of the Bredons necks, adapted from the information board described in the text. The two lava splines are aligned along a fissure and are exposed by erosion. (© Pierre Lavina.)

plateau of Margeride, built from basement gneisses and granite.

Descend to the village and continue eastwards along the road towards Viragues. In 2 km, stop again, at the houses of Faufouilloux.

6. *Diatomite of Falfouilloux*   A large deep excavation in whitish materials is seen on the left-hand side, overlain by a glacial horizon. Piles of

129

debris in the Celite quarry warrant inspection since the diatomite deposit is rich in plant impressions (leaves or pollen of swamp cypress, chestnut and *Carya*). It is of Villafranchian age (Pliocene: between 2.0 and 2.5 million years ago) – the time when the last mastodons died out and the horse, elephant and ox began to thrive in this part of Europe.

Return westwards along the D 139 and rejoin the D 680. Now drive through the houses of Chevade, with the crags of la Veissière rising out of the hillside to the north. Cross the Col d'Entemont and, in about 1 km, take a track up to the right.

7. *Ordanchite quarries near le Chaumeil* The track leads to the quarries where the sombre rock has been used for centuries to make the familiar *lauzes* on the roofs of the region's houses. Some analcite occurs in these outcrops, but not the bluish mineral **haüyne**, which, however, occurs nearby (see stop 8).

8. *Knoll at Entremont-haut* If the desire to see some blue haüyne is overwhelming, return to the road and drive back towards Murat for about 1 km, then turn left up a narrow road to Entremont-haut. Park here and seek permission from the farmhouse to walk up onto the prominent knoll (1254 m) where outcrops of haüyne-bearing ordanchite can be inspected. Furthermore, the view from the summit is very fine: even the Sancy massif can be seen in the distance.

Return along the D 680 to Murat and take the D 926 through the town.

9. *Bredons Neck quarry* Leave the town by turning south along the D 926, then, immediately before the petrol station, take a narrow road up to the right (signposted to "Eglise XI siècle de Bredons"). The neck of Bredons rises steeply up ahead and can be seen to be extremely well jointed. Towards the top of the first hill a quarry face can be seen on the right-hand side. Regrettably, at the time of writing, this was closed off and could not be reached on foot; nevertheless, the style and scale of the columnar jointing is clearly seen from the road. So also is a prominent dyke, which cuts the neck.

Continue up to the village and take a right turn towards the church. Park on the flat area in front of the church building.

10. *Summit of Bredons Neck* A footpath climbs steeply up to the summit of the neck. The view from here is superb and the line of necks is well seen, as is the intersection of the valleys once occupied by the valley glaciers of the Alagnon, Chevade, Benêt and Lagnon. By walking close to the edge on the south side, it is possible to view the columnar jointing at the top of the quarry face.

Return to the base of the hill and down the road. Just past the second house, park on the right. (If a minibus is used, walk to 7 and 8 from the church parking spot, as there is not room to park a large vehicle).

11. *Outcrops on south side of Bredons Neck* Opposite the parking spot is a small footpath that leads towards a quarry face somewhat overhung by vegetation. Immediately on the left is a reddish breccia; similar material outcrops along the path, which then is closed off.

12. *Columnar jointing in Bredons quarry face* Walk down the road a little farther until you can find a track just beyond the last house on the left-hand side. Although one cannot reach the quarry face from here, by walking as far as is possible you can find another good view of the basanite neck and its excellent columnar jointing.

Retrace the route to the village (where there is an old stone fountain) and descend to the main D926 road. Turn right and begin the drive up the long hill (direction of St-Flour). In a short while a parking bay and Parc-des-Volcans information board is found on the left-hand side; from here there is an excellent view over the surrounding area. Continue up the hill for about another 400 m and park opposite to a dirt track, which leaves the main road on the right-hand side. Beware of fast-moving traffic along this stretch of road.

13. *Volcanic sequence beside the D926, near milepost "km 4"* Roadside cuttings expose two flows of columnar jointed basalt lava, each with a reddish weathered horizon above. Red cinders cut across the top of one of the flows, these in turn being overlain by a sill. Opposite the two large laybys farther along the road, the sequence is cut by a basalt dyke. The outcrop is an interesting section through the core of a small volcano.

Continue up the hill and follow the road along the top of the lava plateau to Ussel. Just after the village, turn left along the D34 through the hamlet of Ribes. A small outcrop of basalt can be seen on the left-hand side of the road just beyond the last house. Continue on to Celles.

14. *Puy de Talizat* The view across towards the cone of Talizat from Celles is excellent. This is one of the largest scoria cones to punctuate the plateau. It can be reached by taking the path that runs up the cone from the north side of Talizat village.

From Celles follow the D40 southwards, passing through Coltines and then dropping down off the plateau towards le Saillant. Just past the minor crossroads, pull off onto a large parking space on the right-hand side. The castle rises up ahead; the gorge lies to the left.

15. *Falls at le Saillant* Cross the road and walk with care towards the edge of the cliff. Take extreme care here, particularly if wet, since the edge is precipitous and there are no barriers to prevent a fall into the chasm. The view across the valley sees the River Ander cascading over the vertiginous basalt cliffs, which are very well jointed. The columns are much larger towards the base than at the top of the flow. The same flow as forms the organ pipes at St-Flour, where the excursion terminates.

Continue along the D 40 to Andelat, turning up left into the village. Just before the houses, there is access into a large quarry; park here.

16. *Quarry at Andelat*   The quarry has exploited a basalt flow that outcrops in the quarry face. This is overlain by Miocene mudstones and sands, which have been incised by a massive pale-coloured laharic breccia. Rejoin the D 40 and drive to St-Flour.

17. *Organ pipe columns, St-Flour*   The upper town is located on the top of the columnar-jointed hawaiite basalt flow (8.8 million years old) that was seen earlier at le Saillant. The impressive organ-pipe columns are well seen beside the main route between the lower and upper towns. It is possible to pull off the road near the big hairpin bend. The lower part of the flow shows a development of horizontal jointing; the upper part has vertical columns.

Excursion 6
St-Flour–Neuvéglise–Tagenac–Cussac–Tanavelle–St-Flour

Introduction
This half-day itinerary t begins with visits to the basement gneisses south of St-Flour, beyond the periphery of the Cantal volcanic complex. Turning northwest through Tagenac, a stop is made at a rather obscure outcrop of basalt-bearing peridotite nodules at Cussac. Climbing up to the top of the plateau, the route then reaches the summit of the old volcano at Tanavelle, whence there is a superb panorama over the surrounding region. Finally the way is back down off the lava plateau to St-Flour.

Starting point
St-Flour.

Itinerary
Leave St-Flour along the D 921 towards Chaudes-Aigues and continue for about 14 km.

1. *Gneiss exposures beside the D 921*   Pass the turnoff to Neuvéglise and continue along the long downgrade. A major section on the right-hand side exposes the strongly foliated gneisses and migmatites of the basement complex. The sequence is much faulted and there is a wide variation in fold plunge from place to place. Over considerable sections the foliation is at a low angle. The gneiss contains both biotite and sillimanite and is cut by deformed dykes in places.

Continue down to the River Truyère at Bridge of Lanau, where it is possible to turn and return up the incline to Neuvéglise. Turn off the

Figure 7.16 Peridotite nodule in quarry at Cussac.

main road, follow the D48 through the village and drive towards Tagenac, at which point turn right and then immediately left along the D56. In about 3 km turn off right and up a narrow road into Cussac.

2. *Peridotite nodules at Cussac church*   Park near the church at the top of the hill. A basalt flow outcrops at various places, sometimes poking through the road and at others appearing in roadside banks. Within this unit are many pale greenish peridotite nodules (Fig. 7.16), mostly very angular and seldom larger than 15 cm across. More nodules can also can be seen in the stone of the adjacent field walls.

    Carry on along the D57 road, following the top of the lava plateau all the time. Soon, a quarry on the left-hand side exposes the basalt again, this time with reasonably well developed columnar jointing. On reaching the main road (D921), turn left, then, after a further 3 km, turn left up the road signposted to Tanavelle.

3. *Panorama from Tanavelle summit*   On the way up to this remote hilltown more basalt quarries are passed. Tanavelle itself is located at the summit of an old volcano, probably one of the larger in this area. The main point of coming here is to take in the fine view over the plateau of St-Flour. "Sweeping" is the best word to describe it.

    Leaving town is more difficult than getting in: there is a plethora of narrow streets, some of which are one way and others dead-enders. The plan is to leave on the D216 to Liozargues, descending along the edge of the basalt plateau in so doing. Join the main D926 near Roffiac and return to St-Flour. However, any road that leads down hill will do.

Excursion 7
Thiézac – local walking tour

Introduction
The walk up to the shrine above the town reaches an excellent viewpoint across the valley and also some outcrops of coarse breccia beside the grotto. The continuation to the waterfall is a pleasant expedition and reaches a bench cut in a porphyritic basalt lava.

Starting point
Centre of town.

Itinerary
Walk up past Hôtel l'Elancèze and bear right at the main junction. Shortly afterwards, signs indicate a cobbled path on the left-hand side, which climbs up steeply through trees.

1. *Grotto at Notre-Dame-de-Consolation* The path reaches Notre-Dame de Consolation, a chapel high up on a rock ledge above the town. It takes about 30 minutes to reach it. Bear left in front of the chapel and continue steeply up for another 5 minutes to reach the grotto that has been cut in the lower part of an outcrop of massive andesite breccia. This is worthy of some attention, since it is found to be a matrix-supported and virtually monolithic breccia. Continuing past the grotto, the path climbs up steeply through rocks before it attains the cross, perched right on the rock tump. This is a good viewpoint over the valley of the Cère.

   It is now possible to continue along the path (marked as GR400 to Confolens) until it joins a minor road near Lescure. The route then follows the road for about 1 km before striking off left across a rocky shoulder to Cascade de Faillitoux.

   Note This same point can be reached via the D 59 from Thiézac and parking at Lasmolineries. In this case, walk the last 1 km to the falls. The walk from Thiézac via the grotto to the falls will take about two hours.

2. *Cascade de Faillitoux* The waterfall splashes over a ledge in an ankaramitic basalt flow that contains large phenocrysts of black pyroxene and greenish olivine. This an outcrop of one of those rare basic lavas that was extruded in the middle of the volcanic cycle.

   It is now possible to return on foot to Thiézac by following the D 59 into the town. About 1.2 km south of Lasmolineries, extensive outcrops of massive breccia rise above the road on the north side. The walk back to town takes about an hour and a half.

3. *Pas du Roc* If, after visiting the grotto (1), there is no inclination to walk all the way to Cascade de Faillitoux, it is possible to return to the road

and turn left along a metalled minor road that eventually deteriorates into a rocky track. This passes through a few dwellings at Laubret and, in about 2 km, you are beneath the crags of Pas du Roc. The cliffs are composed of massive volcanic breccias, with occasional dykes.

When, almost out of the trees, a cross is seen, return to Thiézac along the same route. This makes a pleasant two-hour saunter, with a fairly stiff climb towards the rocks at the farthest point of the excursion.

Note If you decide to walk down from Col du Pertus (see Excursion 2, pp. 111–118) – a splendid Cantalian excursion on a fine day – it is the GR400 path that you would follow down the mountain. This passes by l'Elancèze before crossing Col de Bellecombe, then turning south towards, and eventually looping back on itself to reach, the Cascade de Faillitoux. It then continues via Confolens and the grotto to Thiézac. This 10 km trip takes a good four hours to complete; however, there is plenty of rock to see on the way and some splendid scenery.

Excursion 8

Aurillac–Miécaze–St-Mamet–Serrières–St-Santin–Decazeville–Aurillac

Introduction

This excursion investigates the major sedimentary basins of Aurillac and Decazeville. The latter represents an arm of the sea that extended into the area during Carboniferous times; the former is of Oligocene age. During the tour you also see tors of basement granite, metamorphosed schists and Carboniferous volcanic rocks.

Starting point
Aurillac.

Itinerary
Leave Aurillac on the N 120 (Argentat). After 11 km, reach the houses of Prentegarde; here turn left along the D 61. In 1.5 km turn again, this time right towards Miécaze. Park on the left before the bridge.

1. *Sands of the Aurillac Basin* The old quarry exposes quartzose sands and muds of Oligocene age, with well developed cross stratification. The structure indicates shallow-water turbulent conditions of deposition.

   Continue ahead through the village and join the D 18. Turn left and follow it until it turns into the D 53 and eventually meets the main N 122. Drive to and park in the town of St-Mamet-la-Salvetat.

2. *Granite of St-Mamet* Walk up the path to the summit of Puy St-Laurent

135

where there is a table d'orientation. This spot provides a fine viewpoint over the granitic basement hills of the Châtaignerie, the Aurillac sedimentary basin and the volcanic complex of the Cantal.

Rejoin the N 122 southwards; at Manhes turn left onto the D 64 towards Serières, stopping near a house on the left-hand side. Walk down towards the river.

3. *Schist/granite contact in Valley of Boussay* Walk down into the constricted valley, whereupon it is possible to observe the passage from mica schists into granite. Down stream, extremely durable schists pass into normal ones and one can find fine andalusite crystals in the more aluminous bands within the former.

4. *Granite tors of Peyrou and la Morétie* Continue south along the D 64 to Boissat, then continue along it to Bonnemayoux and, eventually, to the hamlet of Peyrou. There are some fine rounded granite tors hereabouts, and they can also be seen at la Morétie and Laroque.

Continue now to Miercolès, then take the D 51 to St-Antoine and then the D 45 all the way to its intersection with the D 963. Follow this south to Pont d'Agres, the bridge over the River Lot.

5. *Carboniferous volcanic rocks of Livinhac area* Turn off along the D 627 towards Livinhac. All of the roadside cliffs along here are composed of Carboniferous volcanic rocks that range in composition from andesite to rhyolite. Nice crystals of hornblende can be found in places, together with veins of both calcite and barytes.

6. *Carboniferous "puddingstones" at Decazeville* Return via the D 21 to the D 963 and continue southwards to Decazeville. At Pont du Bourran the basement granites give way to Carboniferous rocks again. In the middle of the long descent there is a sharp left-hand turn. Immediately after this, pull off on the right. Outcrops of conglomeratic sediments ("puddingstone") abound here, and contain rounded pebbles of quartz, volcanic debris and schist. Deposition in a turbulent sedimentary environment is indicated.

Enter Decazeville and follow the road to Aubin (D 221).

7. *Opencast pits at Decazeville* The pits lie on the east side of the road and have exploited a 80 m-thick coal horizon that has the form of an anticline. Current production is about 1100 tonnes per year.

Finally, in the town of Decazeville is a good small geology museum. It is open between 1 July and 15 September, and during French school holidays, between 14.30 h and 18.30 h.

Chapter 8

Geology of Haute-Loire

Although Haute-Loire is peripheral to the region that is the core of this guide, many visitors may take time out to visit the ancient town of Puy-en-Velay. For this reason a brief description of the geology and two brief itineraries are included here.

Haute-Loire has wide open spaces, fine religious architecture and one unique town, le Puy-en-Velay. The town is a delightful surprise – this is a place like no other. The best way to approach is from St-Flour, for, after many miles of wide open countryside moulded from the granitic rocks of the basement and flat-lying lavas, suddenly a valley vista opens up, punctuated by steep rocky spires crowned with bizarre buildings and surrounded by the town's ancient houses and churches. It is truly amazing.

The region is etched largely out of a sequence of flat-lying Pliocene lavas, which range in composition from basalt through andesite to phonolite. Necks and spines punctuate this sequence, the most spectacular being those of basanitic breccia, which rise within le Puy itself. These lavas lie between uplifted horsts of basement granite, gneiss and schist, and are succeeded by a few basaltic scoria cones and valley-bottom flows of Quaternary age. There are also Pliocene sedimentary rocks, which include sands containing the remains of mastodon, early elephant and horse, and some younger Quaternary sedimentary rocks that are largely of periglacial origin. In many respects the trough of le Puy resembles the down-faulted valley of Limagne (pp. 22–24), so we end on a similar note.

The le Puy Basin began its life when, after the Alpine orogeny had shaken up, fractured and generally disturbed the region, the Vellave plateau partially collapsed, forming a downfaulted basin. Erosion of the surrounding mountains produced copious amounts of sediment that partly filled the basin. Into this Tertiary sedimentary sequence was cut the gorge-like valley of the River Loire. However, at the close of Tertiary times, a major volcanic episode overran the area, forcing the course of the Loire to be shifted eastwards. Subsequently, during the Quaternary era, further erosion etched out the volcanic rocks, including the spectacular necks that now are a feature of le Puy-en-Velay.

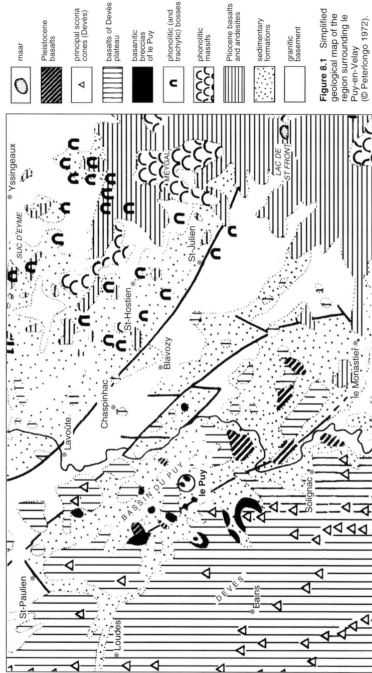

Figure 8.1 Simplified geological map of the region surrounding le Puy-en-Velay (© Peterlongo 1972).

maar

Pleistocene basalts

principal scoria cones (Devès)

basalts of Devès plateau

basanitic breccias of le Puy

phonolitic (and trachytic) bosses

phonolitic massifs

Pliocene basalts and andesites

sedimentary formations

granitic basement

Yssingeaux

SUC D'EYME

St-Hostien

MEYGAL

St-Julien

Chaspinhac

Blavozy

Lavoûte

LAC DE ST-FRONT

le Monastier

St-Paulien

BASSIN DU PUY

le Puy

Solignac

Loudes

DEVÈS

Bains

Because this guide concentrates on Auvergne, herein I will concentrate solely on the rocks of the le Puy Basin, which include basanitic breccias, lacustrine deposits, maars and Strombolian basalt cones (Fig. 8.1). To the west of this basin, and joining it with the eastern edge of the Cantal region, is the volcanic chain of the Devès, a wild region dotted with many Pliocene cinder cones and extensive basaltic lavas that overlie the faulted granitic basement. The journey from St-Flour takes the traveller through the Devès. To the east of le Puy are the Pliocene complexes of Meygal and Mézenc, which at lower stratigraphic levels comprise domes of alkaline trachyte and minor basalts, succeeded by andesite flows and labaradorite; these, in turn, are followed by late-stage trachytes and phonolites, many of which were extruded during violent Peléan eruptions.

The le Puy Basin sequence is of Villafranchian age, this fact having been established largely on the evidence of the lenticles of sandy sedimentary rocks, containing abundant remains of mastodon, which are interleaved with the lavas and breccias. The distinctive breccias occur as volcanic necks, as in le Puy itself, but also form escarpments and sheets. Their fabric and mineralogical composition – the matrices are often very finely pulverized – imply interaction between magma and groundwater below the ground, that is, an origin in hydromagmatism. The lavas are typically undersaturated and alkaline, being particularly rich in soda. Many contain the pyroxene, aegerine. They contrast nicely with those typical of the rest of Auvergne. Geologists continue to debate the degree to which this magmatic series developed by contamination of mantle-derived magmas with continental crust; there is still no consensus.

The Quaternary activity, largely Strombolian in type, followed on from that of the Tertiary, apparently without a hiatus. The several cones of basalt cinders and scoriae are accompanied by many valley-filling flows and are largely confined to the le Puy Basin.

There are two tours in this section; one being a walking trip around the town of le Puy itself and the other a wide-ranging vehicle excursion focusing largely on the area around le Puy and to the River Allier, to the south. Together these provide an introduction to the geology of this interesting but peripheral area.

Excursion 1
Walking tour around le Puy-en-Velay

Introduction
The town is located in a large depression with a slightly convex floor, across which runs a tributary of the Loire (the Borne). To the south, the River Allier flows along a southeast–northwest line. Out of this bowl rise spectacular volcanic rocks that form the focus of this walking tour, which, done at a leisurely pace, takes about a half day, largely on account of the considerable climbing that has to be done.

Starting point
Cathédrale de Notre-Dame.

Itinerary
The fine cathedral is well worth a close inspection, whether or not geology is the main purpose of the visit. Climb the steps on the northern side of the old quarter of town and, via the Route des Tables, to the top of the basaltic edifice upon which it was built. At least an hour is needed to take in the various architectural and ecclesiastical treasures. Having imbibed the delights of this mainly eleventh-century building, take the signed path from the north side and ascend steeply to the top of Rocher Corneille.

1. *Rocher Corneille* The path up to the summit of this jointed volcanic pillar is steep and the journey tiring, but on a fine day is well worth the effort for a superb view over the town and the basin of le Puy. There is a table d'orientation at the summit and a hideous Victorian statue, which was erected in 1860 and painted a repugnant shade of pink. This is best ignored, but not the view, as the panorama provides a fine introduction to the landscapes of the surrounding region – the le Puy Basin encircled by its steep rim and cut through by the River Loire, along a roughly north–south line. The rock itself is an eroded volcanic neck composed of dark basanitic breccia (Fig. 8.2).

    Descend to the west side of the cathedral and, from the cloisters, join the Route de J. de l'Evêque, which passes the twelfth-century Chapelle St-Clair.

Figure 8.2 The silhouette of Roche Corneille framed by trees in the town of le Puy. It is composed of basanitic breccias.

2. *Rocher St-Michel*. Just beyond the church begin the ascent of the challenging 268 steps that lead to the summit of this impressive volcanic pinnacle. Crowning the rock is a fine late eleventh-century chapel (entry fee FF4). There is some splendid black and white masonry adorning the facade and a fine view over the lower town, cathedral and Rocher Corneille. The rock is similar in type to that of (1) above, with a very fine matrix and a slightly different jointing pattern.

Descend to the town and enjoy a well earned rest.

Excursion 2
*Le Puy–Orgues d'Espaly–Mont Denise–le Marais–Polignac–Monistrol–
St-Privat–Polignac–Chaspinhac–le Puy*

Introduction
This excursion provides a general introduction to the geological diversity of the Bassin du Puy and the pleasant countryside of this part of Haute-Loire. It begins and ends in le Puy-en-Velay and would take the best part of a day to complete.

Starting point
Le Puy-en-Velay.

Itinerary
Leave town along the N102, shortly turning right along the D13 towards Blanzac, then, 500 m farther on, bear left towards la Malouteyre. Pass through the village and stop near the obvious bend from which there is a fine view towards the castle of Polignac.

1. *Basanitic breccias near la Malouteyre* A little farther along the road, basanitic breccias are exposed in a small cliff beside the route. These pyroclastic rocks occupy a position at the base of the nearby volcano of la Denise. By climbing up to the left (towards a small reservoir) the breccias are seen to be overlain by a well stratified deposit that cross-cuts the older sequence and dips towards the south. The lower part of the sequence is very finely grained and is the product of breakdown of the breccias and also of the underlying marls; the upper part is enriched in basalt blocks. Together these comprise the Malouteyre Tuffs, which at various locations, together with scoriae from the volcano of la Denise, are interbedded with basanitic breccias.

Continue along the road as far as the TV relay station.

2. *Exposures beside Malouteyre road* The same sequence is exposed along this section of the route, with the scoriae of la Denise inclined to the bedding in the breccias. The morphology of the breccia landscape is nicely seen from here, and behind can be seen the basement granite horst of Chaspinhac.

Return along the same route to the main N102 Clermont-Ferrand road and take the signposted turning to the Orgues d'Espaly.

3. *Orgues d'Espaly* This feature is etched out of a basalt dyke – the remnant of a basalt lava lake – characterized by vertical cooling columns in the interior but inclined ones towards the edge. From its summit there is a particularly fine panorama towards le Puy-en-Velay, with its two prominent volcanic spines, the basaltic plateau beyond and the phonolite hills of Mt Mézenc towards the east. The road then passes along the western slope of the volcano of la Denise, where there is a large quarry cut in the reddened scoriae of this Strombolian cone. A little farther along the route, the basanitic breccias re-appear.

Continue along the main road, passing through Chaspuzac and St-Jean-de-Nay. From here and westwards are good exposures of basement gneiss with augen structures. Continue for a further 3 km and turn left along the D40. In Lapeyre take a minor road on the right towards le Marais-de-Limagne (this section can be either driven or completed on foot).

4. *Quarry en route for le Marais-de-Limagne*   In about 700 m there is a quarry on the left-hand side, with good exposures of well stratified cinders with blocks of peridotite, schist and granite.

5. *Maar crater of le Marais-de-Limagne*   By continuing along the minor road for a further 250 m, you descend towards the floor of a maar crater. On the rim both lapilli and explosion breccias can be found.

   Now return to the D 40 and drive south along it through Beyssac and les Passeyres. Just past the latter hamlet, turn off along the D 25 to St-Privat d'Allier. On reaching the D 589 turn right along it as far as Pratclaux.

6. *Mica schists at Pratclaux.* Along this section of the road are exposures of mica schist and augen gneisses with both muscovite and biotite. These form the basement in the region.

   In Pratclaux cross the River Allier by the narrow bridge and enter Monistrol d'Allier.

7. *Columnar basalts of Monistrol d'Allier.* The river here runs through a gorge and there are good exposures of columnar-jointed basalts. Where exposures allow, these can be seen to rest on pebble beds.

   Now follow the D 301 eastwards to St-Privat d'Allier. Take the turn-off to le Villard.

8. *Basalts of le Villard.* The little hill is composed of surprisingly fresh basalt, which contains abundant but small nodules of greenish perido-tite.

   Rejoin the D 589 and continue to Bains; here, turn left and drive cross country via the D 212 and D 906 to Chaspuzac. In about 5 km take a left turn onto the D 112 and cross to the main N 102 road. Turn right and shortly afterwards left and enter Polignac. Climb up the road to the church, and park.

9. *Roche de Polignac.* Go on foot to the base of the north side of the hill where there are inclined beds of basanitic breccia. As you walk towards the west the slivers of breccia become nearly vertical, suggesting that the Roche was at least in part intrusive in origin.

   Returning to the church, note that the surround to the porch is built from the same breccias. Also from the village there is a fine vista towards the breccia rock of Bilhac, lying to the north-northeast. Now take the minor route north towards Rochelimagne (D 13), then bear right along the D 253 to Chambeyrac.

10. *Rochelimagne flow and the neck of la Roche.* On the left will be seen the small basaltic flow of Rochelimagne, and ahead the upstanding scoria neck of la Roche can be seen, which is of Pleistocene age. To the east of the basalt and separated from it by a small depression are Oligocene marls, which are overlain by a periglacial **solifluction** deposit.

Return to the D25 and drive east towards St-Vincent. This route crosses another basalt flow, this one emanating from Mont Courant to the north, and then follows the surface of the granitic horst of Chaspinhac. At the first major bend on the descent into St-Vincent, stop for the fine view of the basin of l'Emblavès. This descent is somewhat similar to that from the Puy-de-Dôme onto the floor of the Limagne Valley.

Now take the main D103 road southwards towards le Puy-en-Velay; this follows the course of the River Loire. The road enters the wooded Gorge de Peyredeyre, whereupon, after emerging from it, the upstanding scoria boss of Mont Serre comes into view, rising up from the edge of the granitic escarpment. Continue back to le Puy-en-Velay.

Chapter 9

Matters non-geological

Auvergne, as we have seen, is an excellent stamping ground for the geologist, but it also has wide appeal for those with an interest in history, architecture, botany, birds, photography, walking or cuisine. Many books have been written about the various facets of the region, some of which are listed in the appendix at the end of this guide. By way of an introduction, here are some observations about the non-geological aspects of Auvergne.

History

Discovery of stone weapons in Auvergne implies that humans lived here probably 15000 years ago; 10000 years later they appear to have developed a somewhat more settled lifestyle, as pottery remains show. The region was then invaded by the Celts at about 800 BC and it was from the Arverne tribe that the region eventually took its name. It seems likely that their first settlements were on the floor of the fertile Allier Valley. Celtic life went on relatively undisturbed – although not without skirmishes with neighbouring tribes – until Fabius Maximus arrived in 121 BC and endeavoured to impose Roman rule. Around 50 BC, the famous Celtic leader, Vercingetorix, had the ambition to liberate Gaul from this Roman stranglehold and had a measure of success before finally being overcome. The Romans prevailed here for several centuries thereafter. The Romans developed spas, agriculture and mining, and built a network of roads across the region. They also constructed potteries, which began to export ceramics to all parts of the Roman Empire, including Britain. It was the Romans who replaced the Celtic centre of Gergovia (whose exact position is in some doubt) with a new one called Augustonemetum, upon which now stands the city of Clermont-Ferrand. However, by around 50 BC their empire had fallen into decay and a period of anarchy ensued; these Dark Ages were pretty dismal in Auvergne. In AD 614 the Duchy of Aquitaine was formed, a self-governing body that encompassed all of the region, but this was badly organized and it eventually collapsed in AD 768 when the

145

first Carolingian king of France – Pippin the Short – took over. He was both corrupt and confused, but his one saving grace was that he was also the father of Charlemagne, who became the charismatic leader of the Franks in the late seventh century AD. This amazing man built a Frankish empire that stretched from Poland to the Pyrénées but, regrettably, it fell apart quickly after he died in AD 814. Aquitaine (which encompassed Auvergne) was handed down to various of Charlemagne's descendants in turn, but these were not powerful people and Auvergne was soon taken over by the Counts of Auvergne, of whom William the Pious was particularly effective. He held territory not only in Auvergne but also in Velay, Lyonnais, Berry and Mâconnais, and bestowed upon himself the grandiose title of Duke of Aquitaine. A period of instability and strife followed, rule being based on something akin to the feudal system prevalent in Britain at the time. A plethora of fortified houses and castles were erected to try and keep things in order, but the economy stagnated and local militia roamed the countryside taking the law into their own hands. By the beginning of the thirteenth century AD, such was the chaos that inroads into the territory had been made by King Philippe Auguste of France; eventually Auvergne became a part of that kingdom. However, conditions remained much the same: there were highwaymen, corrupt barons, brigands and hired assassins, which all conspired to make life for the peasant of Auvergne something like a hell on Earth.

By the fifteenth century, things had begun to improve and a modicum of peace had returned to the region. As a result the economy took a turn for the better and towns such as Thiers, le Puy-en-Velay and Ambert began to enjoy a measure of prosperity. King Louis XI's various expeditions into the region then slowly eroded the power of the corrupt barons, but it took almost another century for the rule of the king to truly hold sway. Progress was then hindered by a religious upheaval, as Calvin's brand of Protestantism made inroads into hitherto Catholic France. The Protestant Huguenots rose up against the Catholic majority, but were ruthlessly slaughtered in 1572, on St Bartholomew's Day. Violent reactions followed until, after the accession of Henry IV in 1593, rights of worship were given to the Huguenots, and religious tensions eased. During the early part of the seventeenth century, Cardinal Richelieu broke the power of many of the corrupt nobles and this happened again with Louis XIV somewhat later. However, it was not until the middle of this century that some measure of stability and prosperity returned to the region. There was then more religious unrest – epitomized by the Camisard Revolt – which ended with the death of the Camisard leader, Roland. Eventually King Louis XVI signed an Act of Tolerance in 1787, and Protestants and Catholics returned to their villages and lived in relative peace – as they have to this day. It was at this time that many of the old castles were destroyed, as the power of the barons was removed once and for all. At the end of the seventeenth century and beginning of the eighteenth, the cereal harvest throughout France failed; as a result, more and more of previously uncultivated Auvergne land was turned into small farms by the peasant farmers. When the French Revolution came, it was met with fairly widespread apathy in the region and certainly there was strong resistance to the military conscription imposed on all the young men by the King of France, since they were sorely needed to work the expanding Auvergnat farms.

The modern history of the region was shaped by all kinds of odd events. One was the discovery by Lord Macintosh that, by dissolving rubber in benzene, one could manufacture a waterproof substance that prevented the rain getting through a coat – the macintosh had been invented. By chance, his niece married a Frenchman, an event that, after a convoluted series of occurrences, led to the opening of the famous Michelin tyre factory in Clermont-Ferrand in the nineteenth century. The railway arrived in the middle of the same century, playing a part in bringing more goods and people into the region but also facilitating an outward movement of younger Auvergnats to cities such as Paris. In time, many of these became successful business people – the Auvergnats were never afraid to work hard – and several rose to the ranks of ministers in the French government. Thus, President Georges Pompidou was born in the Cantal, Valérie Giscard d'Estaing is distantly related to one of the old noble families of Auvergne, and Jacques Chirac derives from Corrèze, on the edge of the region. Today Auvergne benefits from much more help from central government than once it enjoyed – could there be a connection here with the origins of certain French politicans? The road system is good, tourism is thriving and Clermont-Ferrand is an industrial centre of some considerable worth. However, rural depopulation has continued, as more young people move to the towns to secure jobs and escape from the harshness of rural hill farming. Nevertheless, the peaceful old villages of the region now welcome thousands of tourists every year and this is bringing a measure of prosperity to this difficult region.

Flora

In May and June the meadows of the region are carpeted with wild flowers. Although not many of these are rare, the sheer abundance of narcissi, cranesbill geraniums, saxifrages, globeflowers, orchids, marsh marigolds and forget-me-nots is quite unforgettable. In autumn – particularly in the Cantal and Haute-Loire – myriad crocuses thrust up through the meadow grass, while the deciduous trees turns wonderful shades of yellow, russet and dark brown.

There are four distinct climatic elements within the region: in the southern part (the southern Cantal) a Mediterranean climate ensures a presence of holm oak and parched scrubland where vines, pomegranates and cacti live. At higher altitudes is the zone of sweet chestnuts, the area of Cantal known as the Châtaignerie being the province in whose valleys these nuts grow. The centre for their marketing is the town of Maurs. A Central European element characterizes the ridges of the Forez, in the east, where silver fir and beech thrive. At higher altitudes in these regions there are stands of Scots pine. An Atlantic characteristic is to be found where cool moist summers and mild wet winters prevail (rain at low altitude and snow at high altitude). The expanses of Auvergnat, heathland with their Spanish gorse bushes and lilac and white Cornish heather, typify this particular niche. The fourth climatic zone is the Boreal, where alpine species dominate.

Thus, in Auvergne's more elevated parts, there are many natural rockeries, delightful to the flower lover. They begin to appear above 1200 m and are best developed on the volcanic mountains of the region. The alpine pastures lie above the forested hillsides and are a treasure house for plants people. Here are pulsatillas, anemones, globeflowers, saxifrages, stonecrops, mountain avens, rock jasmine, marsh gentians, Solomon's seal, narcissi, daffodils, lilies, alpine violas and pansies. In spring, as the snows retreat, crocuses, snowdrops and jonquils appear. Then, of course, there are the fungi. Their sequence of appearance during a typical year is: morel, fairy-ring champignon, chanterelle, cep, saffron milk cap, parasol mushroom, wood blewit and blue leg blewit.

Birds, butterflies and moths

Auvergne is not well known as a bird-watcher's region. However, it does have some worthwhile places, particularly in the Cantal. Some observers consider St-Flour as one of the best bases for this occupation and, in any case, its location atop a thick lava remnant makes it an attractive town with a commanding view over the surrounding plains and hills. North of the town is Col de Fageole (1104 m) and this place, in particular, is a good spot to observe raptors such as red kite, Montague's harrier, booted eagle and buzzard. Another well known raptor area is near Lempdes, where riverside fields are bustling with smaller birds such as cirl bunting, redbacked shrike, wryneck and others of the smaller passerines.

Butterflies and moths are abundant in the region. The alpine and

Mediterranean habitats are of particular interest to the northern European lowlander and species of both types have infiltrated the periphery of the Auvergne. Among the rarer representatives are the poplar admiral, the cranberry fritillary and heath fritillary, and various species of ringlets have been recorded. In the damp meadows around Mont-Dore the violet copper is found. In addition to these less common types there are abundant swallowtails and many smaller species such as the skippers. As to the moths, there are wild silk moths, the death's head hawk moth, and the humming-bird hawk moth is also to be found alongside many smaller species. The lepidopterist will not be at all disappointed.

Cuisine

Auvergne is not counted among the great gastronomic regions of France, yet eating here is an entirely satisfactory experience, particularly on account of the fine cheeses, fungi, and hearty soups and stews. Of the cheeses, the better known are Cantal, St-Nectaire, Bleu d'Auvergne and Fourme d'Ambert. However, there are many other delicious types, many made from goat's milk, one I particularly like being the peppery Jaberon. Although not made in the region, Roquefort is also to be found here. Honey is also widely made and can be bought in towns, at farms and along the roadside in many places (this is also true of the cheeses). Country cooking is of a high standard and hotel kitchens produce a wide variety of delicious Auvergnat food. Lentils, cheese, cabbage, potatoes, garlic, onions and rabbit all feature widely in regional dishes. Dishes such as le Truffade are a particular favourite of mine – rich young Cantal cheese, onions, potatoes and bacon lardons all cooked into a delicious cholesterol-rich cake. Coq au vin is also widely served, and tripoux (bits of sheep's feet stuffed and folded up in pieces of the stomach) is much tastier than it sounds. Fish, particularly trout, is widely available and is always fresh from the local rivers. Fruits such as plums, cherries and myrtles are often used in desserts, particularly tarts and flans. Auvergnat wines are not widely renowned, yet there are several that are well worth seeking out. Côtes d'Auvergne, made from the Gamay grape, has a good nose and fresh bouquet. The best cru délimités are Châteaugay, Chanturgue, Madargue and Boudes. Good rosé wine crus are those of Corent, Châteaugay and Boudes. Red, rosé and white wines from the vineyards around St-Pourçain are also generally of good quality and are widely available in the region. Local herbal liqueurs and aperitifs are also sold in local shops, especially delicatessen in the more touristic villages.

Appendix

Viewpoints and information boards

PUY-DE-DÔME
- *Summit of Puy de Dôme* Viewpoints from many places on the summit plateau giving panoramic views over a very wide region. Also two informative tables d'orientation.
- *Calvaire, Châtelguyon* Lava butte within town with an old castle at the summit. Fine panoramic views over Chaîne des Puys, Forez and Limagne. Table d'orientation.
- *Château de Tournoël, near Volvic* Excellent panorama over the downfaulted valley of Limagne towards the Forez mountains.
- *Gour de Tazenat* Excellent views towards Puy de Dôme from the rim of the maar. Good Information board on west rim.
- *Roadside pullout on the D 90 between Orcines and Ternant* Superb view over Limagne Valley.
- *Footpath between le Cheix and Old Royat* Fine views over Limagne and city of Clermont-Ferrand.
- *Summit of Puy de Côme* Excellent panorama over Chaîne des Puys.
- *Summit of Puy de Pariou* Superb vista over adjacent puys.
- *Information board beside the D 941b* Between Col des Goules and la Fontaine du Berger.
- *Information board in layby beside the D 89 between Randanne and Col de la Ventouse* Informative display regarding the formation of Puy de la Vache, Puy Lassolas etc. Good view of southern end of Chaîne des Puys.
- *Veyre-Monton* Magnificent panoramic view over Allier Valley from the summit of the hill on which this town nestles. Ghastly marble statue with good fossils.
- *Gergovie summit* Wonderful views

over the Limagne towards the Forez. Nearby bar/café.
- *Summits of Puy de la Vache and Puy Lassolas* Excellent views of southern end of Chaîne des Puys, towards Massif du Monts-Dore.
- *Château-Rocher, Gorges de la Sioule* Fine view over meander in river below.

SANCY
- *Information board in layby beside the D 5 north of Murol* Fine view over Château de Murol and inverted relief of Puy de Bessolles. Information board regarding geology and wildlife.
- *Summit of Dent du Marais* Fine view over Lac Chambon and Château de Murol, etc.
- *Viewpoint beside footpath above the D 5 between Murol and Bessolles (west flank of Puy de Bessolles)* Views towards Murol, Chambon-sur-Lac and Col de la Croix Morand.
- *Roche Romaine* Summit rocks give good view over surrounding region.
- *Summit of Puy de Montchal, Lac Pavin* View towards Sancy massif.
- *Summit of Puy de Sancy* Table d'orientation at summit and magnificent vista. It is said one can see Mont Blanc on a clear day. Explanatory geology board en route to summit from top cable car station.
- *Le Salon du Capucin* Good view over glaciated Monts-Dore Valley and Grande Cascade.
- *Grande Cascade – top of descent footpath* Good views down over the Monts-Dore Valley.
- *Col de Guéry* Large car-park and information board for view down Orcival Valley and the upstanding necks of Roches Tuilière and Sanadoire.
- *Roche Sanadoire* Small car-park and information board beside the D 9.

- Information board and viewpoint for Roche Sanadoire and neighbouring Tuilière.
- *Summit of Roche Sanadoire* Not for the faint-hearted. Good viewpoint.
- *Summit of Banne d'Ordanche* Panorama over the surrounding region.
- *Rocher de l'Aigle* Superb view up and down Chaudefour Valley.
- *Chaudefour Valley* Information board beside the D 637 with useful cross section and information about Chaudefour Valley.
- *Souilles-et-Combrailles* Information board beside the D 987.

CANTAL AREA
- *Bort-les-Orgues* Information board for Gorge de la Dordogne in picnic area beside the D 679 above town.
- *Bort-les-Orgues* Table d'orientation near summit reached by turnoff from the D 979. Good vantage point over reservoir and organ pipe lava remnant.
- *D 922 near la Pradelle* Parking area with excellent information board regarding Cantal region.
- *Roc du Merle* Viewpoint immediately below rock beside the D 680.
- *Summit of Puy Mary* Magnificent panorama and table d'orientation for Cantal.
- *Mandailles Valley* Information board for Puy Griou group of necks beside the D 17.
- *Alquier* Table d'orientation can be reached by walking up hill above lonely farm. Excellent vantage point and informative board.
- *Salers* Good view over the surrounding countryside from the viewing platform at the upper end of the town.
- *Col de Curebourse* Good view over the glaciated valley of the Cère.
- *Summit of Bredons Neck* Fine view over aligned necks and Murat town.
- *Information board beside the D 926 between Murat and St-Flour* A large layby and information board dealing with the geology of the Albepierre-Bredons Necks is to be found here. There is also a good view.
- *St-Flour upper town* There is a good view down over the surrounding valley and lava plateau from near

the cathedral.
- *Tanavelle summit* Wide panorama over the surrounding region from the summit of this old volcano.
- *Summit of Puy de Peyre-Arse* Superb panorama.
- *Information board and viewpoint beside the D 67 between la Lioran and Lastache* Information "Font du Cère" regarding agriculture in Cère Valley.
- *Summit of Plomb du Cantal* Excellent viewpoint with table d'orientation.

LE PUY-EN-VELAY AREA
- *Rocher Corneille* Magnificent view and information board.
- *Rocher St-Michel* Another good viewpoint with table d'orientation.
- *Roadside viewpoint beside the N 102 on approach to le Puy from northwest* Superb panorama.
- *Summit of Polignac town* Viewpoint over surrounding countryside.

Visitor and information centres

There are some extremely well prepared information centres which explain – albeit in French – the various aspects of the Auvergne: geology, ecology, agriculture, wildlife, flora, history etc. They also keep a stock of leaflets, books, postcards, maps and some sell videos.
- **Maison des Volcans**, Montlosier, near Randanne, 63210 Rochefort-Montagne, Puy-de-Dôme. This complex houses the administrative offices of the Volcanoes Park (tel. 73 65 67 19). Opening hours: Every day except Saturday and Sunday, from 8.30 to 12.00 and from 13.30 to 17.30.
- The same complex also houses an information centre, with permanent and changing exhibitions, including videos. Outside is a picnic area and children's adventure playground with a volcanoes theme. No refreshment facilities, but toilets and water fountains available (tel. 73 65 67 19). Opening hours: From 15 June to 30 September, every day from 10.00 to 12.30 and from 14.30 to 19.00.
- **Summit of Puy de Dôme.** Information centre of the Volcanoes Park. This houses permanent exhibits, including

videos, has a tourist and produce shop, bar/restaurant and excellent book and map store. Wonderful situation on summit plateau of the volcano (tel. 73 91 42 46). Opening hours: From 15 June to 30 September, every day from 10.00 to 12.30 and from 14.30 to 19.00.

• **Maison des Volcans**, 10, rue du Président Delzons, Aurillac (Cantal). This houses a permanent information centre for the Volcanoes Park, as well as an exhibition and book store (tel. 71 48 68 68). Opening hours: Every day except Monday morning, Saturday afternoon and Sunday and bank holidays. In July and August, from 10.00 to 12.00 and from 14.00 to 19.00; rest of the year between 8.30 and 12.00 and from 14.00 to 17.30. Entry fee FF10.

• **Maison de la Pierre**, Volvic. An underground exhibit with video presentation, located inside an old lava excavation. Attached is a small but good bookstall, photographic exhibit and postcard sales booth. Toilets and picnic area adjacent (tel. 73 33 56 92.) Opening hours: Open every day except Tuesdays from 15 March to 1 May, and from 30 September to 15 November. Visits run at 10.00, 10.45, 11.30, 14.15, 15.45, 16.30, 17.15 and 18.00. Reservations can be made for groups (discount). French commentary only but English language notes available.

Towns and villages: hotels & campsites

The region has many hotels of a wide range of grades, many bed and breakfast establishments (chambres d'hôtes), and a fair number of campsites. Here are a few recommendations.

Besse-en-Chandesse *Hôtel du Levant* (tel. 73 79 50 17). Reasonable small hotel in town centre. Modern rooms and good restaurant.

Boussac-Bourg *Camping le Château de Poinsouze* (tel. 55 65 02 21). Excellent spacious campsite in grounds of old château and beside lovely lake.

Clermont-Ferrand *Hôtel Arverne-Mercure*, 16, place Delille (tel. 73 91 92 06). No frills, moderate priced hotel

near old town centre. Convenient access and reasonable food.

Chambon-sur-Lac *Hôtel le Grillon* (tel. 73 88 60 66). Excellent small hotel right beside the lake. Comfortable and with fine food. Family run.

Ebreuil *Camping le Filature* (tel. 70 90 71 01). English run campsite beside River Sioule. Very good facilities; shop and bar; rooms to rent.

Giou-de-Mamou *Auberge "la Rocade"* (tel. 71 63 49 18). Small inn set well back off main Aurillac road with excellent restaurant.

Lanobre *Centre Touristique Municipal de la Siauve* (tel. 71 40 31 85). Large well equipped campsite close to Bort-les-Orgues and lake. Large pitches and shop.

Murol *Hôtel de Paris* (tel. 73 88 60 09). Same owners as Hôtel le Grillon. Old fashioned French hotel with excellent food but steep stairs. *Camping la Ribeyre* (tel. 73 88 68 41). Small quiet site beside plan d'eau. Pizzeria, good facilities and near Murol.

Orcines *Hôtel Relais-des-Puys*, la Baraque (tel. 73 62 10 51). Family-run hotel with excellent dining room. In sight of Puy de Dôme.

Polminhac *Hôtel les Parasols* (tel. 71 47 40 10). Roadside hostelry with no frills but clean rooms and basic facilities. Good food.

Royat *Hôtel le Châtel* (tel. 73 35 82 78). Well appointed small hotel with good restaurant. Just below Old Royat.

St-Flour *Hôtel du Nord* (tel. 71 60 28 00). Right in the upper town. Small, very clean and tidy. Excellent restaurant. Parking nearby.

St-Jacques-des-Blats *Hôtel des Chazes* (tel. 71 47 05 68). Good small hotel, off main road and with good restaurant.

Thiezac *Hôtel l'Elancèze* (tel. 71 47 00 22). In centre of town, nice rooms and very good restaurant. Has rooms in nearby annexe.

APPENDIX

Chemical analyses of igneous rocks

	Sancy					Cantal				Puy-de-Dôme		
	1	2	3	4	5	6	7	8	9	10	11	12
SiO_2	43.60	42.20	71.95	59.60	59.76	47.76	45.31	54.30	56.34	47.20	47.05	65.20
TiO_2	2.70	2.85	–	0.20	0.80	2.71	2.79	0.28	0.06	3.06	3.62	0.41
Al_2O_3	13.90	11.60	14.20	20.53	19.08	15.75	14.44	21.91	17.76	16.43	15.40	18.35
Fe_2O_3	8.60	6.25	0.56	1.71	2.25	5.05	5.07	2.64	3.58	5.24	4.56	1.69
FeO	5.10	7.05	1.13	0.66	1.35	6.28	6.83	0.97	0.06	6.50	7.18	1.55
MnO	0.18	0.13	–	0.11	–	0.25	0.18	0.19	0.18	0.17	0.16	0.22
MgO	8.40	12.10	1.12	0.25	0.59	6.45	8.77	0.47	0.40	6.16	6.54	0.25
CaO	11.60	11.90	0.80	1.10	3.16	10.63	11.10	3.46	2.39	9.92	9.89	2.24
Na_2O	3.70	2.65	5.80	6.91	5.69	3.17	3.15	8.13	9.88	3.05	2.42	5.54
K_2O	1.80	1.80	4.42	6.00	5.01	1.35	1.61	5.65	4.97	1.71	2.12	3.86
P_2O_5	0.90	0.70	0.83	1.63	1.63	–	–	0.09	3.19	–	0.89	0.07
H_2O^+	0.25	0.70	0.83	1.63	1.63	–	–	0.09	3.19	–	0.13	0.37
	100.73	100.93	100.81	100.02	100.27	100.00	100.00	98.18	98.16	100.00	99.96	99.75

Key:
1. Olivine-poor basalt, Cascade du Lac de Guéry
2. Olivine-rich basalt, lower flow, Charlannes
3. Rhyolite, middle funicular station, la Bourboule to Charlannes
4. Phonolite, le Piton
5. Phonolite, Roche Sanadoire
6. Average of saturated basalts
7. Average of undersaturated basalts
8. Miaskitic phonolite, Vinsac
9. Agpaitic phonolite, Repastils.
10. Average basalt of Chaîne des Puys
11. Essexibasalt, Puy de Dôme
12. Domite, Puy de Dôme summit

154

Glossary

aa   Term of Hawaiian origin for lava flow with a blocky surface.

airfall deposit   A volcanic rock laid down after the ballistic ejection of material from an eruption.

alkali basalt   Basalt containing normative nepheline. Average silica content 48 per cent.

anatexis   The metamorphic process whereby existing rocks are so changed that they begin to melt.

andesite   Volcanic rock containing pyroxene, intermediate plagioclase and smaller amounts of biotite, hornblende and alkali feldspar. Average silica content 57 per cent.

ankaramite   Cumulate basic lava enriched in pyroxene and olivine or both.

ankaratrite   Biotite-bearing olivine melanephelenite.

aplite   A fine-grain granitic rock.

apophysis   A small subsidiary lobe of magma emanating from the main body.

ashflow deposit   Any volcanic rock formed from the activity of a fast-moving gas-laden flow of lava and rock.

basalt   Plagioclase–clinopyroxene volcanic rock. Average silica content 48 per cent.

basanite   Olivine basalt in which feldspathoids make up more than 10 per cent of the felsic minerals.

benmoreite   Mugearite–trachyte with anorthoclase and normative plagioclase. $Na_2O > K_2O$.

charnockite   A hypersthene-bearing granitic rock formed under dry crustal conditions in continental regions.

cheire   French word for a prominent flow of lava.

cumulodome   Volcanic dome with convex profile that has accumulated by heaping up of viscous magma.

diapir   A podlike body of magma that has risen from a zone of melting towards the surface of the Earth.

diorite   An igneous rock of intermediate silication, containing between 52 and 66 per cent silica.

domite   Biotite–trachyte to trachyandesite. Tridmyite and cristobalite in groundmass.

doréite   Olivine-bearing trachyandesite.

enclave   Inclusion of foreign rock within an intrusive body.

erratic   A boulder left by a glacier after it has melted. Such may have been transported many hundreds of miles from its source region and can provide useful information on ice movement.

essexitic gabbro/basalt   Feldspathoidal monzo-diorite and monzo-gabbro.

evaporite   Deposit formed by evaporation of standing water, e.g. gypsum, anhydrite.

fiammé   Flattened lava fragments, usually deformed after being entrained in pyroclastic flows.

glowing avalanche   A fast-flowing debris flow consisting of molten lava, rock fragments and hot gas. Synonymous with pyroclastic flow.

gneiss   A metamorphic rock of coarse grain size and foliated appearance, formed by regional metamorphism of mainly continental sedimentary and igneous materials.

graben   A fault trough bounded by a pair of normal faults.

grade   The degree to which an existing rock is changed by metamorphism. The higher the temperature and pressure experienced, the higher the metamorphic grade.

granite A crystalline coarse-grain igneous rock of high silication, whose essential minerals are quartz, alkali and plagioclase feldspars, with or without micas and hornblende.

groundmass The finely crystallized material between the phenocrysts in an igneous rock.

hawaiite A mildly undersaturated alkaline volcanic rock.

hydridization The mixing of two or more magma types to produce a mixed rock.

hydromagmatic Style of volcanism involving interaction between hot magma and water (formerly called "phreatomagmatic").

horst An upfaulted block bounded by normal faults.

igneous Rock formed by crystallization from a magma.

ignimbrite A term used to describe pyroclastic flow deposits. Where molten lava fragments have been entrained in such flows, they take on a characteristic flattened fabric and are called **fiammé**.

indurated Hardened by compaction or cementation.

labradorite Used by the French to describe a form of leucobasalt. More widely, the term is applied more specifically to Ca-rich plagioclase feldspar.

laccolith Mushroom-shape igneous intrusion formed by arching of rocks by magma pressure.

laharic Generated by a lahar or mudflow.

latite Volcanic equivalent of monzonite (andesine, alkali feldspar, clinopyroxene rocks with or without olivine and hornblende).

leucocratic Light in colour, i.e. an igneous rock composed of light-coloured felsic minerals.

maar Explosion crater with low rim formed by hydromagmatic eruption.

magma Melted rock material.

mantle That part of the Earth between the crust and the core. It extends from the base of the former (10–80 km down) to about 2900 km.

marl A calcareous mudrock.

matrix The very fine-grain clay and silt particles between the larger grains in a sedimentary rock.

melanocratic Dark in colour, i.e. an igneous rock composed of dark-coloured mafic minerals.

metamorphic facies The sum of the heat and pressure to which a sequence of rocks has been metamorphosed. In order of increasing temperature and pressure, the sequence in regional metamorphism is: zeolite—greenschist—amphibolite—granulite—eclogite.

metamorphism The process of changing rocks deep within the Earth by heat, pressure and directed stress.

migmatite A mixed rock, consisting of both igneous and metamorphic components.

mode [of a rock] The modal composition of an igneous rock is the tally of the minerals visible in that rock, expressed as a percentage.

Mohorovičić discontinuity Seismic horizon where there is an abrupt change in the velocity of earthquake waves. This marks the boundary between the crust and the mantle.

monzonite An igneous rock intermediate in composition between a syenite and diorite. Usually it contains quartz, plagioclase feldspar and alkali feldspar.

moraine A deposit of blocks and sediment left by a glacier upon its retreat.

mugearite Olivine leucotrachyandesite.

neutral buoyancy zone The level to which magma will rise to be in equilibrium between magma pressure and hydrostatic pressure in surrounding rocks.

normative The rock **norm** of an igneous rock is a set of minerals, calculated from its chemical analysis, which would be expected to occur if normal crystallization conditions applied.

ordanchite Leucocratic haüyne trachyandesite.

orogenic belt An elongated body of

rocks, usually located along the margins of a convergent plate boundary, where pressure, heat and burial will eventually generate a new fold mountain chain.

orogeny A phase of mountain building. It is the sum of tectonic processes in which large belts of the Earth's crust are folded, thrust, metamorphosed and intruded by igneous bodies.

oversaturated [igneous rock] Igneous rock in which there is an excess of silica over the other metallic elements present, whereupon quartz occurs as a mineral in the sample.

pahoehoe Term of Hawaiian origin for lava flow with ropy surface.

peneplain A flat landscape surface formed by a lengthy period of erosion.

peridotite Titanaugite–olivine-bearing cumulate rocks. Average silica content 41 per cent.

phonolite Alkali-feldspar rock, often with minor oligoclase and sometimes quartz. Average silica content 59 per cent.

polymict Containing clasts of different rock types (applied to sedimentary rocks).

porphyroblast A large crystal that has grown in a metamorphic rock because of burial and recrystallization.

pyroclastic flow A fast-moving ground-hugging mixture of molten lava, rock fragments and hot gas. Also known as a "glowing avalanche".

pumice Frothed silicic lava, that is, molten rock in which rapid expansion of dissolved gases has led to formation of a product full of gas cavities. Such rocks will float on water.

puy French word for dome- or cone-shaped volcanic eminence.

puzzolane Local term for granulated cinders typical of Strombolian cones of Puy-de-Dôme.

regional metamorphism The kind of metamorphic activity associated with the formation of mountain belts along collisional plate boundaries. It involves recrystallization of pre-existing rocks, impression of new (foliated) fabrics and dewatering of sedimentary rocks.

retrogressive [metamorphism] Meta-

morphism is usually progressive, that is, a series of rocks are changed as temperature and pressure increase with depth of burial or heating up during an orogeny. However, rocks that have been metamorphosed once may be affected a second time (or more times), but not to such a high degree. In this case, new minerals formed during the first phase may be changed to other ones stable at the lesser temperature and pressure. These effects are said to be retrogressive.

rhyolite Volcanic equivalent of granite. Average silica content 72 per cent.

sancyite Leucotrachyandesite, containing tridymite.

saturated [igneous rock] A rock in which the proportions of silica versus metallic elements exactly balances out, so that free quartz does not occur.

schist Foliated metamorphic rock formed by regional metamorphism of mainly sedimentary materials.

scoriae Cinder-like fragments of lava.

sedimentary rock Rock formed from particles of pre-existing rocks (sediment) which have been either cemented or compressed into a coherent body.

stratovolcano A steep-sided volcanic edifice built from both lava and pyroclastic debris. Such volcanoes have a complex history and usually are located along convergent plate boundaries.

subduction The consumption back inside the Earth of ocean floor and associated sediments at convergent plate boundaries. Such recycling happens along inclined **subduction zones**.

syenite An igneous rock of intermediate silication in which alkali feldspar is more abundant than plagioclase.

tephrite Alkali basaltic rock composed of plagioclase, augite and feldspathoid.

thermoluminescence A dating technique in which materials that emit light when heated are exposed to ionizing radiation. Some of the electrons freed by the ionization may be trapped at points of imperfection in the crystal lattice and will remain there at room temperature. Heating of the crystals releases the electrons, which manifest their surplus energy as visible light. The amount of light emitted is proportional to the radiation dose absorbed by the material; thus an age may be established.

tillite   Glacial deposit composed of a mixture of badly sorted blocks and silty material.

trachyte   Alkali-feldspar-rich rock, often with some oligoclase. It may contain either quartz or nepheline. Average silica content 59 per cent.

trachyandesite   Equivalent to a latite with plagioclase (< An 50 per cent).

undersaturated [igneous rock]   Igneous rock in which there is insufficient silica present to combine with the metallic elements and combine with them to produce "normal" feldspar. Instead, some or all of the feldspar is substituted by feldspathoids, which use up less silica.

volcaniclastic rock   Any rock formed via volcanic eruption which involves the fragmentation of either or both of molten lava or pre-existing solid materials. These may be laid down as **airfall** deposits or may be water lain.

Bibliography, related reading and websites

Popular publications

Auvergne/The Rhone Valley, 2nd edn [Michelin Green Guide]. ISBN: 2-06-130402-8 1997. Michelin Travel Publications
France [Insight Guide], 1998 (ISBN 962-421-435-2). A beautifully presented and lavishly illustrated guide to all aspects of visiting France.
Up close France, Fodor, 1998 (ISBN 0-679-03397-1). A good general up-to-date guide to all aspects of visiting France.
Auvergne and Rhône Valley [Michelin Green Guide] (Pneu Michelin, 1998, ISBN 2-06-130402-8). A fund of information on all aspects of the region, with excellent maps, up-to-date information about opening hours, fees and telephone numbers.
Auvergne and the Massif Central, Rex Grizell (Christopher Helm, 1989, ISBN 0-7470-1220-2). A well written book about all aspects of the region, with some lovely photographs and line drawings. A mine of information and, although some of the information may not now be entirely up-to-date, this does not detract from a book which is a delight to read.
Walks in the Auvergne [Footpaths of Europe] (Robertson McCarta, 1989, ISBN 1-85365-140-0). A guide to 400 km of footpaths throughout Auvergne, written in English and with clearly annotated maps.
History, people and places in Auvergne, A. N. Brangham (Spur Books, 1977, ISBN 0-904978-X). Although an old book – and out of print – this is well worth seeking out for its perspectives on all aspects of life in the region. There is much about the history, the nature of the people and anecdotes and facts about the towns and villages.
Histoire géologique du Cantal, H. Bril (La Maison des Volcans, Aurillac, 1987). A locally produced geological guide, in French, with six itineraries in Cantal. Has a useful bibliography of academic publications dealing with this region.
Volcanologie de la Chaîne des Puys, A. de Goër de Hervé & G. Camus (eds) (Parc Naturel Régional des Volcans d'Auvergne, 1991). In French, but there is now an English language version. This handy pocket-sized compendium consists of a colour geological map of the Puy-de-Dôme region and an illustrated geological description in a plastic folder. It has an extensive bibliography (mainly of French publications). It is well prepared and beautifully illustrated, and is obtainable at Maisons des Volcans, the Information Centre on the summit of Puy de Dôme, Maison de la Pierre, and some bookshops.
Promenades à pied dans les volcans d'Auvergne, A. de Bussac (Editions De Borée, Lyon, 1991, ISBN 2-908592-07-X). A handy-sized pocket guide to walks in the Chaîne des Puys region, with annotated maps, nice colour photographs and geological explanations (in French.
Les sites volcaniques d'Auvergne, G. Joberton (Editions De Borée, Cournon d'Auvergne, 1996). Another pocket-size geological guide which includes Haute-Loire. [In French]

Leaflets, brochures and local maps

Auvergne: hotels. Booklet published by Comité Regional du Tourism.
L'Auvergne à la carte. Local hoteliers' leaflet published at Châtelguyon (tel.

73 86 16 52).

Routes des Cotes d'Auvergne. Leaflet on wines published and available locally.
Routes des fromages AOC *d'Auvergne.* Leaflet published by Association des Fromages d'Auvergne, Aurillac and Clermont-Ferrand.
Saint Nectaire. Informative tourist booklet published locally (tel. 73 88 50 86).
Vallée de la Sioule (English version). Tourist map. Published by SMAT Val de Sioule, Mairie, 03450 Ebreuil (tel. 70 90 78 30).
Volcanoes of Auvergne Regional Park. Leaflet published by Volcanoes Park headquarters at Montlosier – 6370 Aydat (tel. 73 65 67 19).

Technical publications

Aubert, H. & G. Camus 1974. Deep structure of Chaines-des-Puys. *Bulletin Volcanologique* **38**, 443–51.
Bout, P. & R. Brousse 1969. Guide to excursion C13: Auvergne-Velay. Congres viii, INQUA, Paris.
Brousse, R. 1960. *Description géologique du massif volcanique du Mont-Dore (partie ouest).* Thesis, Université de Paris.
Glangeaud, Ph. 1969. Livret-guide de l'excursion a9: Massif Central. VIIIe Congrés de l'Union Internationale pour l'étude du Quaternaire, Paris.
de Goër de Hervé, A. 1977. Structure et dynamique des edifices volcaniques recents du Massif central. *Symposium J. Jung,* Plein Air Service, Clermont-Ferrand, pp. 345–76.
Michel, R. 1953. Contribution a l'étude petrographie des pépérites et du volcanism tertiare de la grande Limagne. Mémoire 5, Société Histoire Naturelle d'Auvergne, Clermont-Ferrand.
Peterlongo, J. M. 1972. *Massif Central* [Guides Geologiques Regionaux]. Paris: Masson.
Sorenson, H. (ed.) 1974. *The alkaline rocks.* Chichester, England: John Wiley.
Villemont, B., J-L. Joron, H. Jaffrezic, M. Treuil, R. C. Maury, R. Brousse 1980. Cristallisation fractionnée d'un magma basaltic alcalin: la serie de la chaine des Puys, II: géochemie. *Bulletin Minéralogique,*
Villemont, B., J-L. Joron, H. Jaffrezic, M. Treuil, R. C. Maury, R. Brousse 1980. Cristallisation fractionnée d'un magma basaltic alcalin: la serie de la chaine des Puys, II: géochemie. *Bulletin Mineralogique* **103**(2), 267–86.
Wimmenaur, W. 1974. The alkaline province of Central Europe and France. In *The alkaline rocks,* H. Sorenson (ed.), 238–70. Chichester, England: John Wiley.

Websites

Altavista.look-smart.com/ This Altavista search engine provides access to a general France bibliography and services search, region-by-region.
www.mont-dore.com/ General information, including some geology, for the Sancy region.
clrwww.in2p3.fr/ Accesses a range of information about Auvergne.
www.obs.univ-bpclermont.fr/ University site dealing with seismology of region.
www.france-hotel-guide.com/ Lists regional hotels.
www.auvergne-thermale.tm.fr/ Deals with the spas and other tourism aspects.
www.auvergnet.com/ Describes the wines of the region.
www.le-guide.com/ General information about the Auvergne region.
www.crt-auvergne.fr/ Tourism in Auvergne.
www.guideweb.com/ Lists some recommended camping sites.

Geological tours to Auvergne

Journeys of Special Scientific Interest In association with Voyages Jules Verne, 21 Dorset Square, London NW1 6QG (tel. 0207 616 1000). Small specialist organization running escorted small-group 7- and 10-day geological holidays to various parts of the world, including Auvergne.

Index of places and names

Index of topics